T0135557

Model Predictive Control for Nonlinear Continuous-Time Systems with and without Time-Delays

Von der Fakultät Konstruktions-, Produktions- und Fahrzeugtechnik
und dem Stuttgart Research Centre for Simulation Technology
der Universität Stuttgart zur Erlangung der Würde eines
Doktor-Ingenieurs (Dr.-Ing.) genehmigte Abhandlung

Vorgelegt von

Marcus Reble

aus Stuttgart

Hauptberichter:	Prof. Dr.-Ing. Frank Allgöwer
Mitberichter:	Prof. Riccardo Scattolini
	Prof. Dr. Lars Grüne

Tag der mündlichen Prüfung: 28. Februar 2013

Institut für Systemtheorie und Regelungstechnik

Universität Stuttgart

2013

Bibliografische Information der Deutschen Nationalbibliothek

Die Deutsche Nationalbibliothek verzeichnet diese Publikation in der
Deutschen Nationalbibliografie; detaillierte bibliografische Daten sind
im Internet über http://dnb.d-nb.de abrufbar.

D 93

ISBN 978-3-8325-3381-6

Logos Verlag Berlin GmbH
Comeniushof, Gubener Str. 47,
10243 Berlin
Tel.: +49 (0)30 42 85 10 90
Fax: +49 (0)30 42 85 10 92
INTERNET: http://www.logos-verlag.de

Für meine Familie

The best way to predict the future is to invent it. — Alan Kay

Acknowledgements

The results reported in this thesis have been developed during my time as research and teaching assistant at the Institute for Systems Theory and Automatic Control at the University of Stuttgart. There are several people to whom I would like to express my gratitude.

First and foremost, I would like to thank my advisor Prof. Dr.-Ing. Frank Allgöwer for all the opportunities he gave me and for inpiring me with his great enthusiasm for control theory. He created a unique environment for research and gave me lots of freedom in my research. Moreover, he provided support for attending several international conferences, where I was able to present my work and get into contact with other researchers in my field.

I want to thank Prof. Riccardo Scattolini and Prof. Dr. Lars Grüne for the interest in my work, the valuable comments, and for being members of my doctoral exam committee.

I also want to thank Prof. Daniel E. Quevedo for making my research stay in Newcastle, NSW, Australia so memorable and productive at the same time, for all the fruitful discussions, and for pushing me in the right directions.

Furthermore, I want to thank the German Research Foundation (DFG) for support of my work in two ways. First, the Priority Programme 1305 "Control Theory of Digitally Networked Dynamical Systems" supported my work financially. Second, the Graduate School Simulation Technology (GS SimTech) within the Stuttgart Research Centre for Simulation Technology supported my research stay in Newcastle, NSW, Australia and gave me outstanding possibilities to discuss with people from different research areas.

I would like to thank my colleagues, both at the IST and from other institutions, the guests at the IST, and the students who I have supervised. All of you contributed to a great atmosphere and it was a pleasure working with you.

Last but not least, I am eternally indebted to my parents for all their support and love.

Steinenbronn, March 2013
Marcus Reble

Table of Contents

List of Symbols

The following list only contains symbols which are used throughout the thesis. Symbols which are defined locally are not listed below.

Sets

$\mathbb{N}$	set of natural numbers
$\mathbb{N}_0 = \mathbb{N} \cup \{0\}$	set of non-negative integers
$\mathbb{R}$	set of real numbers
$[c_1, c_2]$	interval $\{k \in \mathbb{R} \mid c_1 \leq k \leq c_2\}$ for constants $c_1, c_2 \in \mathbb{R}$
$[c_1, c_2[$	interval $\{k \in \mathbb{R} \mid c_1 \leq k < c_2\}$ for constants $c_1, c_2 \in \mathbb{R}$
$\mathbb{P}_{\geq c}$	set $\{k \in \mathbb{P} \mid k \geq c\}$ for $\mathbb{P} \subseteq \mathbb{R}$ and for a constant $c \in \mathbb{R}$
$\mathbb{P}_{\leq c}$	set $\{k \in \mathbb{P} \mid k \leq c\}$ for $\mathbb{P} \subseteq \mathbb{R}$ and for a constant $c \in \mathbb{R}$
$\mathbb{P}_{> c}$	set $\{k \in \mathbb{P} \mid k > c\}$ for $\mathbb{P} \subseteq \mathbb{R}$ and for a constant $c \in \mathbb{R}$
$\mathbb{P}_{< c}$	set $\{k \in \mathbb{P} \mid k < c\}$ for $\mathbb{P} \subseteq \mathbb{R}$ and for a constant $c \in \mathbb{R}$
$\mathbb{R}^n$	n-dimensional Euclidean space
$\mathcal{PC}(\mathbb{A}, \mathbb{B})$	set of all piece-wise continuous functions $\varphi : \mathbb{R} \supseteq \mathbb{A} \to \mathbb{B}$
$\mathcal{C}_\tau = \mathcal{C}([-\tau, 0], \mathbb{R}^n)$	Banach space of continuous functions mapping the interval $[-\tau, 0] \subset \mathbb{R}$ into $\mathbb{R}^n$
$\mathcal{K}$	a function $\alpha : \mathbb{R}_{\geq 0} \to \mathbb{R}_{\geq 0}$ is said to belong to class $\mathcal{K}$ $(\alpha \in \mathcal{K})$ if it is continuous, $\alpha(0) = 0$, and strictly increasing
$\mathcal{K}_\infty$	a function $\alpha : \mathbb{R}_{\geq 0} \to \mathbb{R}_{\geq 0}$ is said to belong to class $\mathcal{K}_\infty$ $(\alpha \in \mathcal{K}_\infty)$ if $\alpha \in \mathcal{K}$ and $\alpha(s) \to \infty$ as $s \to \infty$

Vectors, Matrices, and Norms

I	identity matrix
0	matrix of zeroes
A^T	transpose of matrix A
$\begin{bmatrix} A & B \\ \star & C \end{bmatrix}$	$\star$ denotes the symmetric part of a matrix, i.e., $\begin{bmatrix} A & B \\ \star & C \end{bmatrix} = \begin{bmatrix} A & B \\ B^T & C \end{bmatrix}$
$A \succ 0, (A \succeq 0)$	matrix A is positive (semi-)definite
$A \prec 0, (A \preceq 0)$	matrix A is negative (semi-)definite
$\lambda_{\min}(A)$	smallest real part of the eigenvalues of matrix A
$\lambda_{\max}(A)$	largest real part of the eigenvalues of matrix A
$|x|$	Euclidian norm of $x \in \mathbb{R}^n$
$\|A\|$	spectral norm of matrix A
$\|x_t\|_\tau$	norm on $\mathcal{C}_\tau$ defined as $\|x_t\|_\tau = \sup_{\theta \in [-\tau, 0]} |x(t + \theta)|$

Functions

$\exp(s)$	exponential function e^s with $e = \lim\limits_{n \to \infty} \left(1 + \frac{1}{n}\right)^n$
$\operatorname{ceil}(s)$	smallest integer larger or equal to $s \in \mathbb{R}$
$\operatorname{floor}(s)$	largest integer less or equal to $s \in \mathbb{R}$
$\operatorname{sign}(s)$	signum function

System Variables and Control Parameters

t	time
x	state
x_0	initial state
u	input
f	vector field
$\mathbb{U}$	input constraint set
T	prediction horizon
δ	sampling time
F	stage cost
E	terminal cost function(al)
Q, R	weighting matrices for a quadratic stage cost
Ω	terminal region

Additional Notation for Time-Delay Systems

τ	time-delay
$x(t)$, $x(t - \tau)$	instantaneous and delayed state
x_t	full state of a nonlinear time-delay system at time t defined by $x_t(s) = x(t + s), s \in [-\tau, 0]$
$\varphi \in \mathcal{C}_\tau$	initial function for a time-delay system
V	Lyapunov-Razumikhin function

Acronyms

CLF	control Lyapunov function(al)
CSTR	continuous stirred tank reactor
FDE	functional differential equation
LMI	linear matrix inequality
MPC	model predictive control
NMPC	nonlinear model predictive control
ODE	ordinary differential equation
RFDE	retarded functional differential equation
TDS	time-delay system

Abstract

Model predictive control (MPC) is a modern control method based on the repeated online solution of a finite horizon optimal control problem. It is particularly attractive due to its ability to take hard constraints and performance criteria directly into account. The objective of this thesis is the development of novel MPC schemes for nonlinear continuous-time systems with and without time-delays in the states which guarantee asymptotic stability of the closed-loop. The most well-studied MPC approaches with guaranteed stability use a control Lyapunov function as terminal cost. Since the actual calculation of such a function can be difficult, it is desirable to replace this assumption by a less restrictive controllability assumption. For discrete-time systems, the latter assumption has been used in the literature for the stability analysis of so-called unconstrained MPC, i.e., MPC without terminal cost and terminal constraints.

The contributions of this thesis are twofold. In the first part, we propose novel MPC schemes with guaranteed stability based on a controllability assumption, whereas we extend different MPC schemes with guaranteed stability to nonlinear time-delay systems in the second part.

In the first part of this thesis, we derive for the first time explicit stability conditions on the prediction horizon as well as performance guarantees for unconstrained MPC for continuous-time systems. Starting from this result, we propose novel alternative MPC formulations based on combinations of the controllability assumption with terminal cost and terminal constraints. Thereby, we show connections of our results to previous MPC schemes and highlight advantages. One of the main contributions is the development of a unifying MPC framework which allows to consider both MPC schemes with terminal cost and terminal constraints as well as unconstrained MPC as limit cases of our framework.

In the second part of this thesis, we show that several MPC schemes with and without terminal constraints can be extended to nonlinear time-delay systems. Due to the infinite-dimensional nature of these systems, the problem is more involved and additional assumptions are required in the controller design. For MPC schemes with terminal constraints, we prove that stability conditions similar to the delay-free case are sufficient for closed-loop stability. However, the calculation of suitable terminal cost functionals and terminal regions based on the Jacobi linearization about the origin is more difficult. We propose and investigate different procedures to overcome these difficulties. For MPC schemes without terminal constraints, we discuss MPC schemes with and without terminal cost functionals. If the terminal region is defined as a sublevel set of the terminal cost, we show that the terminal constraint can be omitted from the optimal control problem while maintaining asymptotic stability. Similar to the results in the first part of the thesis, explicit stability conditions on the prediction horizon are derived based on a modified controllability assumption suitable for time-delay systems.

Deutsche Kurzfassung

Modellprädiktive Regelung nichtlinearer zeitkontinuierlicher Systeme mit und ohne Totzeiten

Motivation und grundsätzliche Fragestellungen

Die modellprädiktive Regelung, im Englischen als *model predictive control* (MPC), *moving horizon control* oder *receding horizon control* bezeichnet, ist ein modernes modellbasiertes Regelungsverfahren und erfährt sowohl in der Grundlagenforschung und Literatur als auch in praktischen Anwendungen erhebliche Beachtung.

Ein wesentlicher Grund für den Erfolg von MPC ist die intuitive zugrunde liegende Idee. Im Gegensatz zu den meisten anderen Regelungsverfahren wird in MPC keine explizite Abbildung der gemessenen Systemzustände auf den Eingang berechnet. Stattdessen ist diese implizit gegeben durch die Lösung eines Optimalsteuerungsproblems basierend auf dem gemessenen Systemzustand und einem Modell des zu regelnden Systems. Hierbei wird zu jedem Abtastzeitpunkt t_i ein Optimalsteuerungsproblem auf einem endlichen Prädiktionshorizont T gelöst und die dabei berechnete optimale Eingangstrajektorie bis zum nächsten Abtastzeitpunkt $t_{i+1} = t_i + \delta$ auf das System angewendet. Durch die wiederholte Anwendung dieses Prinzips über jeweils verschobene Prädiktionshorizonte wird ein geschlossener Regelkreis erreicht. Die Grundidee der modellprädiktiven Regelung ist in Abbildung 1 veranschaulicht.

Die weiteren Vorteile der modellprädiktiven Regelung gegenüber herkömmlichen Regelungsverfahren sind vielfältig. So kann die Einhaltung harter Eingangs- und Zustandsbeschränkungen garantiert werden, es ist prinzipiell möglich nichtlineare Mehrgrößensysteme zu betrachten und gewünschte Kriterien für eine hohe Regelgüte können explizit im Reglerentwurf berücksichtigt werden.

Für eine weitergehende Einführung in MPC und einen umfassenden Überblick über die theoretischen Ergebnisse in diesem Bereich verweisen wir auf die Übersichtsartikel (Findeisen et al., 2003; Magni and Scattolini, 2004; Mayne et al., 2000) und die Bücher (Camacho and Bordons, 2004; Goodwin et al., 2005; Grüne and Pannek, 2011; Maciejowski, 2002; Rawlings and Mayne, 2009).

Die vorliegende Arbeit beschäftigt sich im Wesentlichen mit einer der grundlegenden Fragestellungen in der Regelungstheorie: Unter welchen Bedingungen ist die nominelle asymptotische Stabilität des geschlossenen Regelkreises gewährleistet? Die optimale Regelung auf einem unendlichen Horizont garantiert unter schwachen Annahmen asymptotische Stabilität des geschlossenen Kreises. Im Gegensatz hierzu ist dies für MPC mit endlichem Prädiktionshorizont im Allgemeinen nicht gewährleistet. Dieser Effekt wurde von Raff et al. (2006) an einem praktischen Beispiel verdeutlicht. Während MPC-Schemata mit Endkosten und Endbeschränkungen, sowie die zugehörigen Stabilitätsbedingungen, als

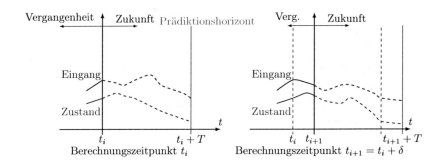

Abbildung 1: Grundidee der modellprädiktiven Regelung.

sehr gut verstanden angesehen werden können (Chen and Allgöwer, 1998; Mayne et al., 2000), so ist dies für das sogenannte *unbeschränkte MPC* (engl.: unconstrained MPC), d.h. MPC ohne Endkosten und Endbeschränkungen, bisher nur für zeitdiskrete Systeme der Fall (Grüne et al., 2010a). Dieses Regelungsverfahren ist insbesondere bedeutsam durch die weitverbreitete Anwendung in der industriellen Praxis und die einfache Formulierung.

Die vorliegende Arbeit leistet wissenschaftliche Beiträge in zwei Bereichen der modellprädiktiven Regelung. Zum einen betrachten wir nichtlineare zeitkontinuierliche Systeme und präsentieren neue MPC-Schemata, für die rigoros Stabilität gewährleistet werden kann. Hierbei stehen Stabilitätsbedingungen basierend auf einer sogenannten *Kontrollierbarkeitsannahme* (engl.: controllability assumption) im Vordergrund. Zum anderen untersuchen wir die Anwendbarkeit von MPC auf nichtlineare Totzeit-Systeme, d.h. Systeme mit verzögerten Zuständen, welche eine wichtige spezielle Klasse unendlich-dimensionaler Systeme darstellen. Hierbei betrachten wir MPC-Schemata sowohl mit als auch ohne Endbeschränkungen und Endkosten. Im Folgenden erläutern wir die Forschungsbeiträge genauer.

Modellprädiktive Regelung für nichtlineare zeitkontinuierliche Systeme

In dem ersten Teil der vorliegenden Arbeit beschäftigen wir uns mit nichtlinearen zeitkontinuierlichen Systemen beschrieben durch gewöhnliche Differentialgleichungen. Die klassischen Stabilitätsbeweise in MPC beruhen auf der Verwendung einer lokalen *Kontroll-Lyapunov-Funktion* (engl.: control Lyapunov function (CLF)) als Endkostenfunktion (Chen and Allgöwer, 1998; Mayne et al., 2000). Die Existenz einer solchen Funktion ist zwar eine gerechtfertigte Annahme, die tatsächliche Berechnung stellt sich allerdings für nichtlineare Systeme im Allgemeinen als sehr schwere oder sogar unlösbare Aufgabe heraus. Daher werden in praktischen Anwendungen oftmals unbeschränkte MPC-Verfahren, d.h. MPC ohne Endbeschränkungen und ohne Endkosten, verwendet. Es ist bekannt, dass diese Verfahren für einen „hinreichend langen" Prädiktionshorizont die Stabilität des geschlossenen Kreises gewährleisten (Jadbabaie and Hauser, 2005). Für zeitdiskrete Systeme existieren darüber hinaus Ergebnisse für die explizite Berechnung eines solchen Prädiktionshorizonts basierend auf einer Kontrollierbarkeitsannahme (Grimm et al., 2005; Grüne, 2009; Grüne et al., 2010a).

Diese Kontrollierbarkeitsannahme verlangt nur eine obere Schranke an die optimalen Kosten des zu lösenden Optimalsteuerungsproblems und ist somit weniger einschränkend als die Kenntnis einer geeigneten CLF. Für zeitkontinuierliche Systeme sind solche expliziten Bedingungen allerdings bisher nicht bekannt.

Ein erster Forschungsbeitrag dieser Arbeit ist die Formulierung geeigneter Stabilitätsbedingungen für unbeschränktes MPC für nichtlineare zeitkontinuierliche Systeme. Darüber hinaus stellen sich unmittelbar weitere Fragen, die bisher in der Literatur nicht beantwortet wurden: Können die bei der Analyse von unbeschränktem MPC angewandten Hilfsmittel auch dazu beitragen, neue Einsichten in die klassischen MPC-Schemata mit Endkosten und Endbeschränkungen zu erlangen? Inwiefern bestehen Verbindungen zwischen den verschiedenen Klassen von modellprädiktiven Regelungsverfahren und lassen sich beide Klassen in einem vereinheitlichten Rahmen betrachten? Unter welchen Bedingungen ist es vorteilhaft, die Annahmen aus beiden Klassen gleichzeitig zu berücksichtigen, d.h. eine Kombination der Kontrollierbarkeitsannahme mit Endkosten und/oder Endbeschränkungen? Im ersten Teil dieser Arbeit beantworten wir diese Fragestellungen.

Modellprädiktive Regelung für nichtlineare Totzeit-Systeme

In dem zweiten Teil dieser Arbeit betrachten wir die Anwendung der modellprädiktiven Regelung auf nichtlineare Totzeit-Systeme beschrieben durch funktionelle Differentialgleichungen. Diese Systeme bilden eine wichtige Klasse unendlich-dimensionaler Systeme und treten in der Modellierung vieler technischer, biologischer und gesellschaftlicher Systeme auf, bei denen die zukünftige Entwicklung nicht nur vom aktuellen Systemzustand abhängt, sondern auch von vergangenen Zuständen. Wichtige Beispiele für solche Systeme beschreiben den Transport von Material und Information, hierbei lassen sich u.a. Rührkesselreaktoren mit Rückführung und die Datenübertragung in geschlossenen Regelkreisen nennen. Nichttechnische Beispiele sind die Glukose-Insulin-Regulierung, die Übertragung von Krankheiten und Preisschwankungen. Für weitere Details und eine umfassende Liste an weiteren Beispielen verweisen wir auf (Kolmanovskii and Myshkis, 1999, Kapitel 2). Durch die unendliche Dimension des Zustands sind sowohl die Stabilitätsanalyse als auch der Reglerentwurf erheblich aufwendiger als für Systeme ohne Totzeiten. Insbesondere gibt es nur wenige Verfahren für den Reglerentwurf, die es erlauben harte Beschränkungen zu berücksichtigen. Dies motiviert die genauere Untersuchung von MPC für diese Systemklasse. Während für lineare Totzeit-Systeme bereits zahlreiche Ergebnisse in der Literatur vorhanden sind, gibt es für nichtlineare Totzeit-Systeme nur sehr eingeschränkte Resultate, die entweder ein globales Kontroll-Lyapunov-Funktional als Endkostenfunktional (Kwon et al., 2001a,b; Mahboobi Esfanjani and Nikravesh, 2009a) oder eine erweiterte Gleichheits-Endbeschränkung (Raff et al., 2007) voraussetzen. Beide genannten Ansätze sind für die praktische Anwendung im Allgemeinen ungeeignet.

Aus diesem Grund betrachten wir alternative MPC-Schemata und leiten Bedingungen her, die asymptotische Stabilität des geschlossenen Kreises rigoros garantieren. Zum einen erweitern wir das klassische Stabilitätsresultat für MPC mit lokalen Endkosten und Ungleichungs-Endbeschränkungen auf nichtlineare Totzeit-Systeme und schlagen Verfahren zur Berechnung geeigneter stabilisierender Reglerparameter vor. Zum anderen berücksichtigen wir MPC-Schemata ohne Endbeschränkungen, sowohl mit Endkosten als auch ohne Endkosten. Die genauere Betrachtung der genannten Fragestellungen bildet den zweiten Teil dieser Arbeit.

Forschungsbeiträge und Gliederung der Arbeit

Die nachfolgende Übersicht zeigt die Gliederung der Dissertation und erläutert die wichtigsten Forschungsbeiträge der einzelnen Kapitel.

Kapitel 2 — Background (Grundlagen)   In diesem Kapitel fassen wir die Grundlagen zusammen, die im Zusammenhang mit den Ergebnissen dieser Arbeit stehen. In Abschnitt 2.1 erläutern wir die Grundidee und bestehende Ergebnisse im Bereich der modellprädiktiven Regelung. In Abschnitt 2.2 beschäftigen wir uns mit den Grundlagen nichtlinearer Totzeit-Systeme, eine spezielle Klasse unendlich-dimensionaler Systeme.

Kapitel 3 — Model Predictive Control for Nonlinear Continuous-Time Systems (Modellprädiktive Regelung für nichtlineare zeitkontinuierliche Systeme)   In diesem Kapitel behandeln wir die modellprädiktive Regelung für nichtlineare zeitkontinuierliche Systeme. Es werden Stabilitätsbedingungen für fünf verschiedene MPC-Schemata hergeleitet, jeweils basierend auf einer Kontrollierbarkeitsannahme. Diese Annahme ist weniger einschränkend als die Kenntnis einer CLF vorauszusetzen und erlaubt explizite Bedingungen an einen stabilisierenden Prädiktionshorizont für unbeschränktes MPC, siehe Abschnitt 3.2. Hiervon ausgehend untersuchen wir Vorteile durch die Verwendung zusätzlicher Gewichtungsterme in den Abschnitten 3.3 und 3.4. Wir zeigen, dass Stabilität für kürzere Prädiktionshorizonte gewährleistet werden kann durch die Verwendung einer Endkostenfunktion, die keine CLF ist, sondern eine ähnliche, aber erheblich abgeschwächte Bedingung erfüllt. Dasselbe Ziel kann durch eine exponentielle Gewichtung der laufenden Kosten erreicht werden. Außerdem betrachten wir mögliche Verknüpfungen der Kontrollierbarkeitsannahme mit Endbeschränkungen in den Abschnitten 3.5 und 3.6. Wir zeigen, dass für geeignet umformulierte Optimalsteuerungsprobleme mit zusätzlichen neuen Gewichtungstermen eine nur lokal vorausgesetzte Kontrollierbarkeitsannahme ausreicht, um Stabilität zu gewährleisten. Darüber hinaus lassen sich sowohl unbeschränktes MPC sowie das klassische MPC-Verfahren mit Endkosten und Endbeschränkungen als Grenzfälle aus dem von uns vorgeschlagenen MPC-Verfahren mit Integral-Endkosten herleiten, wie in Abschnitt 3.5 beschrieben. Dies ist insbesondere von Bedeutung, da beide bestehenden MPC-Verfahren bisher stets getrennt in der Literatur untersucht wurden.

Die wichtigsten Forschungsbeiträge dieses Kapitels sind:

- Wir leiten Stabilitätsbedingungen und Abschätzungen der Regelgüte für unbeschränktes MPC für nichtlineare zeitkontinuierliche Systeme her. Diese basieren auf einer Kontrollierbarkeitsannahme und der Lösung eines zugehörigen unendlich-dimensionalen Optimierungsproblems.

- Wir geben Stabilitätsbedingungen für unbeschränktes MPC mit zusätzlicher positiv semi-definiter Endkostenfunktion, die keine CLF sein muss, an.

- Für den Sonderfall einer exponentiellen Kontrollierbarkeitsannahme und der Verwendung eines zusätzlichen exponentiellen Gewichtes werden analytische Ausdrücke für die Stabilitätsbedingungen berechnet.

- Wir schlagen zwei MPC-Schemata mit Endbeschränkungen vor, bei denen Stabilität auch gewährleistet werden kann, falls die Kontrollierbarkeitsannahme nur in einer lokalen Umgebung der Ruhelage erfüllt ist.

- Wir zeigen anhand der in dieser Arbeit untersuchten MPC-Schemata Verbindungen zwischen unbeschränktem MPC und MPC mit Endkosten und Endbeschränkungen auf.

Teile dieses Kapitels basieren auf Reble and Allgöwer (2011, 2012b); Reble et al. (2012a,b).

Kapitel 4 — Model Predictive Control for Nonlinear Time-Delay Systems (Modellprädiktive Regelung für nichtlineare Totzeit-Systeme) In diesem Kapitel betrachten wir verschiedene MPC-Schemata für nichtlineare Totzeit-Systeme mit garantierter Stabilität des geschlossenen Regelkreises. Hierbei werden MPC-Schemata sowohl mit als auch ohne Endkosten und Endbeschränkungen untersucht. Die Stabilitätsbedingungen für MPC mit Endkosten und Endbeschränkungen in Abschnitt 4.2 sind konzeptionell den bekannten Ergebnissen für Systeme ohne Totzeiten sehr ähnlich, die tatsächliche Berechnung der Reglerparameter stellt sich allerdings bedingt durch die unendliche Dimension des Zustandsraums als erheblich schwieriger heraus. Vier Verfahren zur Berechnung geeigneter Endkosten und Endregionen basierend auf der Jacobi-Linearisierung um den Ursprung werden in Abschnitt 4.3 vorgestellt. Wir beweisen Stabilitätsbedingungen für MPC-Verfahren ohne Endbeschränkungen in Abschnitt 4.4 mit Endkosten und in Abschnitt 4.5 ohne Endkosten. Hierbei wird im ersten Fall ein Endkostenfunktional und eine Endregion, die als Unterniveaumenge der Endkosten definiert ist, verwendet. Das Optimalsteuerungsproblem umfasst jedoch keine Endbeschränkung, da die Endregion nur zur Stabilitätsanalyse verwendet wird, jedoch nicht im eigentlichen Regler. Im zweiten Fall wird eine Kontrollierbarkeitsannahme angewendet ähnlich zu der Annahme, die in Kapitel 3 für Systeme ohne Totzeiten betrachtet wurde.
Die wichtigsten Forschungsbeiträge dieses Kapitels sind:

- Wir erweitern die klassischen Stabilitätsbedingungen für MPC mit Endkosten und Endbeschränkungen auf den Fall von nichtlinearen Totzeit-Systemen.

- Wir entwickeln vier Verfahren zur Berechnung des Endkostenfunktionals und der Endregion basierend auf der Jacobi-Linearisierung um den Ursprung und formulieren jeweils hinreichende Bedingungen in Form von linearen Matrixungleichungen (engl.: linear matrix inequalities (LMIs)). Die Eigenschaften sowie Vor- und Nachteile der Verfahren werden erläutert.

- Wir geben zwei MPC-Schemata ohne Endbeschränkungen sowie zugehörige Stabilitätsbedingungen an.

- Wir vergleichen die in diesem Kapitel vorgeschlagenen MPC-Verfahren anhand zweier numerischer Beispiele: ein akademisches Beispiel, welches im gesamten Kapitel mehrmals zur Verdeutlichung herangezogen wird, sowie das Modell eines kontinuierlich betriebenen Rührkesselreaktors mit Rückführung.

Die Ergebnisse zu MPC mit Endbeschränkungen innerhalb dieses Kapitels basieren auf Mahboobi Esfanjani et al. (2009); Reble and Allgöwer (2010a,b, 2012a); Reble et al. (2011b), während die Resultate zu MPC ohne Endbeschränkungen teilweise auf Reble et al. (2011a,c) basieren.

Kapitel 5 — Conclusions (Fazit)   In diesem Kapitel fassen wir die in dieser Arbeit erzielten Ergebnisse zusammen und setzen diese in einen größeren Zusammenhang. Des Weiteren zeigen wir mögliche Fragestellungen für zukünftige Forschungsarbeiten auf.

Chapter 1

Introduction

1.1 Motivation and Focus of the Thesis

Model predictive control (MPC) is a modern model-based control method relying on the repeated online solution of a finite horizon optimal control problem. It has received remarkable attention in academic research and is widely used in practical applications. The reasons for its great success are manifold: MPC is able to guarantee the satisfaction of hard input and/or state constraints, it can deal with nonlinear systems with multiple inputs, and a performance criterion can be directly taken into account in the control setup. In the last decades, significant progress has been made regarding practical implementation – starting from early applications to rather slow systems in the process industry and going to considerably faster dynamics such as in the area of automotive engineering. Furthermore, a solid theoretical foundation is available with rigorous and well-understood stability proofs for MPC schemes which use a local control Lyapunov function and terminal constraints. However, these theoretically well-founded MPC schemes are typically not used in practical applications. Hence, there is still an even more substantial gap between theory and practice than for most other modern control disciplines. This motivates the research on rigorous stability conditions for practically relevant MPC schemes as well as novel alternative MPC formulations for different classes of system.

While the existence of a local control Lyapunov function is a reasonable assumption, the actual calculation of this function is in general a challenging task for nonlinear systems. For this reason, the so-called *unconstrained MPC* scheme, i.e., MPC without terminal cost function and without terminal constraint, enjoys widespread use in industrial applications. Hence, the investigation of unconstrained MPC is of particular interest in order to bridge the gap between theory and practice. For this MPC scheme, it is well-known that closed-loop stability is not guaranteed in general, but only for a prediction horizon chosen "large enough". For discrete-time systems, explicit conditions on the length of the prediction horizon are available in the literature based on a so-called *controllability assumption*. The aforementioned controllability assumption only requires an upper bound on the optimal cost. This assumption and concepts from relaxed dynamic programming allow to remove the more restrictive assumption of a local control Lyapunov function in the proof of closed-loop stability. For continuous-time systems, however, similar explicit stability conditions in terms of the prediction horizon are not available in the literature. Moreover, several other questions are still without answer. For instance, can the analysis tools used in the stability proof of unconstrained MPC give new insight into the classical MPC schemes with terminal cost and terminal constraints? What are the connections between the different MPC schemes with seemingly different stability analysis and is there some kind of "unifying framework"? Moreover, is it beneficial to combine the controllability assumption from unconstrained MPC with terminal cost terms and/or terminal constraints, and if yes, in which way should this combination be made? In the first part of this thesis, we give answers to these questions.

In the second part of this thesis, we are concerned with a slightly different set of questions. Instead of deriving novel MPC schemes, we are interested in extending existing stability conditions for different MPC schemes to more general classes of systems. More specifically, we consider nonlinear time-delay systems (TDS) in this thesis, i.e., systems with a time-delay in the states, a special class of infinite-dimensional systems. Time-delay systems, also called differential-difference equations or systems with aftereffect or dead-time, describe many practically relevant processes, for instance when the transport of material and/or data is considered. Due to their infinite-dimensional nature, stability analysis and controller design becomes more difficult even for linear time-delay systems. Unsurprisingly, most available control methods for this class of systems do not allow to take hard input constraints into account. For this reason, we investigate different MPC schemes with and without terminal constraints and/or terminal cost terms and derive novel stability conditions. Furthermore, we pay particular attention to the calculation of the involved control design parameters.

In conclusion, the objective of this thesis is twofold. First, we derive novel stability conditions based on a so-called *controllability assumption* for MPC schemes for finite-dimensional nonlinear continuous-time systems. Second, we investigate the use of model predictive control for *nonlinear time-delay systems*. In the following section, we explain the results obtained in this thesis in more detail.

1.2 Outline and Contributions of the Thesis

The outline and the contributions of the thesis are as follows.

Chapter 2 — Background In this chapter, we give a brief overview of the most important existing results related to the work in this thesis. In Section 2.1, we summarize previous results on model predictive control for finite-dimensional continuous-time systems. Section 2.2 introduces time-delay systems, a particular class of infinite-dimensional systems, and recalls important properties as well as previous work.

Chapter 3 — Model Predictive Control for Nonlinear Continuous-Time Systems In this chapter, we consider model predictive control for continuous-time systems. We present novel stability conditions for five MPC schemes based on a controllability assumption, which is less restrictive than the assumption of a local control Lyapunov function. Starting from unconstrained MPC in Section 3.2, we investigate possible benefits of additional weighting terms in Sections 3.3 and 3.4. We can guarantee stability for shorter prediction horizons by using a terminal cost function, which does not need to be a CLF, but satisfies a relaxed similar condition. Similarly, an exponential weighting on the stage cost also allows to reduce the stabilizing prediction horizon. Moreover, we investigate for the first time the combination of the controllability assumption with terminal constraints in Sections 3.5 and 3.6. For an appropriately defined new optimal control problem, a local controllability assumption is sufficient for stability and we recover both unconstrained MPC and classical MPC with terminal cost and terminal constraints as limit cases of our MPC setup with integral terminal cost in Section 3.5.

The main contributions of this chapter are:

- We derive stability conditions and performance estimates for unconstrained MPC for nonlinear continuous-time systems based on a controllability assumption and a corresponding infinite-dimensional optimization problem.

- We derive stability conditions for unconstrained MPC with an additional positive semi-definite terminal cost, which does not need to be a control Lyapunov function.

- We provide stability conditions for the special case of an exponential controllability assumption and the use of an additional exponential weighting on the stage cost.

- We propose two novel MPC setups with guaranteed stability based on only local controllability assumptions in combination with terminal constraints.

- We discuss the connection of these results to previous results on MPC with terminal constraints.

In this chapter, we do not present a separate example section. Instead, we rather use several illustrating examples throughout the different sections. Parts of this chapter are based on Reble and Allgöwer (2011, 2012b); Reble et al. (2012a,b).

Chapter 4 — Model Predictive Control for Nonlinear Time-Delay Systems In this chapter, we propose different MPC schemes for nonlinear time-delay systems with guaranteed asymptotic stability. The results include MPC schemes with and without terminal cost terms and/or terminal constraints. While the stability conditions for MPC with terminal cost terms and terminal constraints in Section 4.2 are conceptually very similar to the well-known results for delay-free systems, the actual calculation of an appropriate terminal cost functional and a controlled invariant terminal region turns out to be significantly more difficult. Four procedures for calculating the terminal cost and terminal constraints based on the Jacobi linearization about the origin are derived in Section 4.3. Stability conditions for unconstrained MPC schemes with and without terminal cost are derived in Sections 4.4 and 4.5, respectively. The first scheme uses a terminal cost functional and a terminal region defined as sublevel set of the terminal cost. The terminal constraint is omitted from the optimal control problem and only used in the stability analysis. The second scheme uses a controllability assumption similar to the one used in Chapter 3 for delay-free systems.

The main contributions of this chapter are:

- We extend classical stability conditions for MPC schemes with terminal cost terms and terminal constraints to nonlinear time-delay systems.

- We provide four different schemes to calculate the terminal cost and terminal region based on the Jacobi linearization about the origin and formulate exemplary conditions in terms of linear matrix inequalities (LMIs) for each scheme. We discuss properties of each scheme and compare the advantages and disadvantages.

- We propose two MPC schemes without terminal constraints for nonlinear time-delay systems with guaranteed stability.

- We compare the presented MPC schemes by evaluating two numerical examples. The first one is a simple academic example used throughout the chapter, the second is the model of a continuous stirred tank reactor with recycle stream.

The results on MPC with terminal constraints presented in this chapter are based on Mahboobi Esfanjani et al. (2009); Reble and Allgöwer (2010a,b, 2012a); Reble et al. (2011b). Parts of the results on MPC without terminal constraints in this chapter are based on Reble et al. (2011a,c).

Chapter 5 — Conclusions In this chapter, we summarize the main results of the thesis and indicate possible directions for future research.

Chapter 2

Background

In this chapter, we provide a brief overview of the most important existing results related to the work in this thesis. In more detail, we present the basic idea of model predictive control and summarize existing results on guaranteed nominal stability in Section 2.1. Section 2.2 gives an introduction to time-delay systems with a particular focus on the definition of stability, stability conditions, and control.

2.1 Model Predictive Control

Model predictive control (MPC) is one of the most successful modern control methods and is based on the repeated solution of an open-loop finite horizon optimal control problem. Compared to classical control methods, it offers several advantages such as the guaranteed satisfaction of hard constraints and the possibility to take performance specifications directly into account. Due to these advantages, it has received much attention in academic research and finds its way into an ever-growing number of practical application areas, assisted by the development of dedicated numerical optimization methods and increasing computing power. In this section, we provide the necessary background of MPC for the remainder of this thesis. After explaining the basic idea of MPC in Section 2.1.1, Section 2.1.2 gives an overview of the existing literature in the field of MPC with a particular focus on nominal stability. Section 2.1.3 presents a well-established general design framework for MPC schemes with guaranteed nominal stability using terminal cost functions and terminal constraints.

2.1.1 Basic Principle of Model Predictive Control

The basic principle underlying model predictive control is rather simple and intuitive. Predictions based on a nominal model of the system and the most recent available measurement of the state of the system are employed to determine the "best" possible control action over a finite time horizon. This optimal control input is applied until new measurement data becomes available. At this time, the procedure is repeated using the new measurement and over a shifted prediction horizon.

Slightly more formally, the principle of MPC can be summarized in the following algorithm, see also Figure 2.1. For a rigorous setup, we refer to Section 2.1.3 and the respective sections in the subsequent chapters.

Algorithm 2.1 (Basic Idea of Model Predictive Control). *For given constant prediction horizon T and sampling time δ, at each sampling instant $t_i = i\delta$, $i \in \mathbb{N}_0$,*

1. *measure the state $x(t_i)$,*

2. *solve an open-loop finite horizon optimal control problem over the time interval $[t_i, t_i + T]$, and*

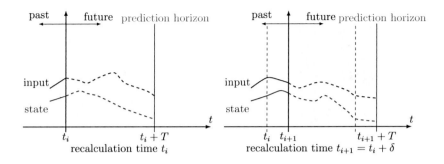

Figure 2.1: Basic principle of model predictive control.

3. *apply the optimal open-loop input to the system until new measurement is available at the next sampling instant* $t_{i+1} = t_i + \delta$.

It is well-known that under weak assumptions optimal control over an infinite prediction horizon guarantees asymptotic stability of the closed-loop. In contrast, MPC with a finite prediction horizon does not ensure stability in general. A simple practical example for instability of the closed-loop is presented by Raff et al. (2006). It is shown that MPC without additional stabilizing constraints can destabilize an open-loop stable four tank system when the prediction horizon is chosen "too short". The question which additional ingredients are necessary in order to guarantee stability received remarkable attention in academic research over the last decades. In the following section, we provide a short overview of the literature in the area of MPC with a particular focus on stability conditions.

2.1.2 Review on Model Predictive Control

In this section, we give a brief overview of several important results in MPC. For a more detailed introduction to MPC and a comprehensive survey of academic research in MPC, we refer to the overview papers (Findeisen et al., 2003; Magni and Scattolini, 2004; Mayne et al., 2000) and books (Camacho and Bordons, 2004; Goodwin et al., 2005; Grüne and Pannek, 2011; Maciejowski, 2002; Rawlings and Mayne, 2009). Different aspects of practical applications of MPC and several examples are reported in (Darby and Nikolaou, 2012; Del Re et al., 2010; Kano and Ogawa, 2010; Qin and Badgwell, 2000, 2003).

The basic idea of using a combination of online optimization and a moving horizon for control was first developed in the 1960s and early 1970s (Lee and Markus, 1967; Nour Eldin, 1971; Propoi, 1963; Rafal and Stevens, 1968; Zadeh and Whalen, 1962). First industrial applications in the process industry are reported in the late 1970s (Cutler and Ramaker, 1980; Richalet et al., 1978). The success of MPC in practical applications led to a significant interest in academic research and the development of a thorough theoretical foundation and rigorous stability guarantees.

The first stability results on MPC with a finite prediction horizon make use of a zero state terminal equality constraint (Alamir and Bornard, 1994; Chen and Shaw, 1982; Keerthi and Gilbert, 1988; Mayne and Michalska, 1989, 1990; Michalska and Mayne, 1991). These results have then been extended towards a terminal inequality constraint, first within the framework of dual-mode nonlinear MPC (Chisci et al., 1996; Michalska and Mayne, 1993; Scokaert et al., 1999) and later towards MPC using control Lyapunov functions (CLF) as terminal cost (Chen, 1997; Chen and Allgöwer, 1998; De Nicolao et al., 1998; Fontes, 2001, 2003; Magni and Scattolini, 2004; Magni et al., 2001; Mayne et al., 2000). These results can be considered as the currently most well-established framework for stability in MPC (Rawlings and Mayne, 2009). For more details, we refer also to Section 2.1.3.

In order to simplify the online computations, several researchers have proposed MPC schemes without additional terminal constraints. One of the earliest works in this direction considers MPC with a finite prediction horizon for linear systems without constraints (Nevistić and Primbs, 1997; Shamma and Xiong, 1997). Another approach is to omit the terminal constraint from the optimal control problem, but to nevertheless guarantee its satisfaction for a defined set of initial states as in (Graichen and Kugi, 2010; Graichen et al., 2010; Hu and Linnemann, 2002; Limon et al., 2003, 2006; Rawlings and Mayne, 2009). By using an appropriate terminal cost, but no terminal constraint, stability conditions are derived in (Alamir and Bornard, 1995; Jadbabaie, 2000; Jadbabaie et al., 1999, 2001b; Parisini and Zoppoli, 1995). These schemes still rely on the use of a control Lyapunov function as terminal cost. In contrast, stability for MPC schemes without terminal cost is shown in (Jadbabaie and Hauser, 2005; Jadbabaie et al., 2001a) for a "sufficiently large" prediction horizon, albeit no explicit bounds on a stabilizing prediction horizon are provided. Such bounds can be derived based on certain controllability assumptions such as the earlier results in (Costa and do Val, 2003; Grimm et al., 2005; Messina, 2006; Tuna et al., 2006). Significantly improved estimates on a minimal stabilizing prediction horizon can be derived by solving a so-called abstract linear program (Giselsson, 2010; Grüne, 2007, 2009; Grüne and Pannek, 2008, 2009, 2010; Grüne and Rantzer, 2006; Grüne et al., 2009a,b, 2010a,b,c; Pannek, 2009; Worthmann, 2012a,b).

The aforementioned references are the foundation for the new results derived in this thesis. It is interesting that, despite the fact that different MPC setups and distinct analysis methods are used, the stability analyses of all these previous and newly developed schemes have one feature in common: The optimal cost function of the finite horizon optimal control problem is used as Lyapunov function for the closed-loop.

However, there also exist several stability results which are not directly related to the well-established framework recalled in Section 2.1.3 and the other results in this thesis. For instance, a substantially different approach was proposed by Michalska (1996), in which the cost function is the sum of the prediction horizon – which is also a decision variable in the optimization problem at each sampling time – and a terminal cost term depending only on the state at the end of the prediction horizon. Another alternative framework for stability is the so-called contractive MPC (Kothare and Morari, 2000; Mejía and Stipanović, 2009; Polak and Yang, 1993a,b; Yang and Polak, 1993). In contrast to using a CLF as terminal cost, several MPC schemes rely on a CLF in a different fashion in order to guarantee stability of the closed-loop. For instance, a global CLF can be applied to design an additional stabilizing constraint in the optimization problem (Mhaskar et al., 2006; Muñoz de la Peña and Christofides, 2008). Similarly, it is possible to consider stabilizing controllers based on a common CLF and to select free design parameters of these controllers employing an optimization over a receding horizon (He et al., 2011; He and Han, 2010). Finally, in the scheme proposed in (Chen, 2010; Chen and Cao, 2012), the stage cost itself is assumed to be a CLF. This allows to guarantee stability for any non-zero prediction horizon. However, these results can be interpreted as a restrictive special case of the aforementioned MPC schemes based on a controllability assumption.

An alternative to the online solution of optimal control problems is given by the so-called explicit MPC schemes. Using parametric programming methods, an explicit optimal control law can be obtained for the case of linear systems (Bemporad et al., 2002a,b; Jones et al., 2007). Approximate solutions with guaranteed properties such as stability have been derived for linear systems (Domahidi et al., 2011; Jones and Morari, 2010; Kvasnica et al., 2011) as well as nonlinear systems (Johansen, 2004; Raimondo et al., 2011; Schulze Darup and Mönnigmann, 2011; Summers et al., 2010), and possible advantages of a combination of explicit MPC and online optimization have been investigated (Zeilinger, 2011; Zeilinger et al., 2011).

There exist several results on MPC for infinite-dimensional systems, although results almost exclusively consider models described by partial differential equations. These results can be distinguished into MPC schemes with terminal cost terms (Dubljevic et al., 2005, 2006a,b; Georges, 2009; Igreja et al., 2011; Ito and Kunisch, 2002; Mohammadi et al., 2010; Ohsumi and Ohtsuka, 2010, 2011; Ou and Schuster, 2010; Pham et al., 2010a,b,c, 2011, 2012; Shang et al., 2007; Tröltzsch and Wachsmuth, 2004; Utz, 2012; Utz et al., 2010) and MPC schemes without terminal constraints and terminal cost terms (Altmüller et al., 2010a,b, 2012; Grüne, 2009). For an overview of MPC for time-delay systems, see Section 2.2.3.

Today, MPC is a thriving field of academic research, highlighted by the large number of publications in journals and international conferences. Current research is dedicated to a huge variety of different areas. One example are new stability results for special system classes such as periodic systems (Böhm et al., 2009; Freuer et al., 2010; Gondhalekar and Jones, 2011; Reble et al., 2009), hybrid systems (Lazar, 2006), and distributed systems (Dunbar and Murray, 2006; Ferrari-Trecate et al., 2009; Keviczky et al., 2006; Müller et al., 2011, 2012; Scattolini, 2009). Other areas are practically motivated control problems such as networked control systems with unreliable communication channels (Findeisen and Varutti, 2009; Grüne et al., 2012; Pin and Parisini, 2011; Quevedo et al., 2011; Reble et al., 2011d, 2012c; Tang and De Silva, 2007) and MPC with economic objectives (Amrit, 2011; Angeli et al., 2012; Diehl et al., 2011; Grüne, 2013). Besides nominal stability of a set-point, other properties are also of interest, such as inherent robustness (Pannocchia et al., 2011; Yu et al., 2011, 2012), tracking (Ferramosca et al., 2009; Limon et al., 2008), and path-following problems (Faulwasser, 2012; Faulwasser and Findeisen, 2009a,b).

2.1.3 Stability in Model Predictive Control

In this section, we recall a well-established general design framework for MPC schemes with guaranteed nominal stability. Similar results can be found in (Chen, 1997; Chen and Allgöwer, 1998; Findeisen et al., 2003; Fontes, 2001; Magni and Scattolini, 2004) for continuous-time systems and in (Magni et al., 2001; Mayne et al., 2000; Rawlings and Mayne, 2009) for discrete-time systems. We consider nonlinear continuous-time systems described by the *ordinary differential equation* (ODE)

$$\dot{x}(t) = f(x(t), u(t)), \quad x(0) = x_0, \tag{2.1}$$

in which $x(t) \in \mathbb{R}^n$ is the state at time t, $x_0 \in \mathbb{R}^n$ is the initial condition, and $u(t) \in \mathbb{R}^m$ is the control input subject to input constraints $u(t) \in \mathbb{U} \subset \mathbb{R}^m$.

The following standard assumptions are used in the following.

Assumption 2.1. *The vector field* $f : \mathbb{R}^n \times \mathbb{R}^m \to \mathbb{R}^n$ *is continuously differentiable and* $f(0,0) = 0$, *i.e.,* $x_s = 0$ *is an equilibrium of system* (2.1) *for* $u_s = 0$.

Assumption 2.2. *System* (2.1) *has a unique solution for all* $t \in \mathbb{R}_{\geq 0}$ *for any initial condition* $x_0 \in \mathbb{R}^n$ *and any piecewise- and right-continuous input function* $u : \mathbb{R}_{\geq 0} \to \mathbb{U}$.

Assumption 2.3. *The input constraint set* $\mathbb{U} \subset \mathbb{R}^m$ *is compact and contains the origin in its interior.*

We use the following standard definitions for stability and asymptotic stability, see for example Khalil (2002); Vidyasagar (1993).

Definition 2.2 (Stability). *The equilibrium* $x_s = 0$ *of system* (2.1) *is called* stable *if for any* $\varepsilon \in \mathbb{R}_{>0}$ *there exists a* $\delta_\varepsilon \in \mathbb{R}_{>0}$ *such that* $|x_0| \leq \delta_\varepsilon$ *implies that* $|x(t)| \leq \varepsilon$ *for all* $t \geq 0$. *It is called* asymptotically stable *if it is stable and there exists a* $\delta_0 \in \mathbb{R}_{>0}$ *such that* $|x_0| \leq \delta_0$ *implies that* $\lim_{t \to \infty} |x(t)| = 0$.

Remark 2.3. *In this work, we sometimes refer to (asymptotic) stability of a system meaning (asymptotic) stability of the equilibrium at the origin.*

The MPC controller is based on the repeated online solution of the following open-loop finite horizon optimal control problem at each sampling time t_i given the measured state $x(t_i)$.

Problem 2.4.

$$\underset{\bar{u} \in \mathcal{PC}([t_i, t_i + T], \mathbb{R}^m)}{\text{minimize}} \quad J_T(x(t_i), \bar{u}) \tag{2.2a}$$

subject to

$$\dot{\bar{x}}(t'; x(t_i), t_i) = f(\bar{x}(t'; x(t_i), t_i), \bar{u}(t')), \qquad t' \in [t_i, t_i + T], \tag{2.2b}$$

$$\bar{x}(t_i; x(t_i), t_i) = x(t_i), \tag{2.2c}$$

$$\bar{u}(t') \in \mathbb{U}, \qquad t' \in [t_i, t_i + T], \tag{2.2d}$$

$$\bar{x}(t_i + T; x(t_i), t_i) \in \Omega, \tag{2.2e}$$

in which

$$J_T(x(t_i), \bar{u}) = \int_{t_i}^{t_i + T} F(\bar{x}(t'; x(t_i), t_i), \bar{u}(t')) \, dt' + E(\bar{x}(t_i + T; x(t_i), t_i)).$$

In this problem, $\bar{x}$ denotes the state predicted over the prediction horizon $T \in \mathbb{R}_{>0}$ based on the model of the system. We assume that the optimal open-loop control which solves Problem 2.4 is given by $u_T^*(t'; x(t_i), t_i)$ for all $t' \in [t_i, t_i + T]$. For a given sampling time $\delta \in \mathbb{R}_{>0}$, the control input to the system is defined by the following algorithm in the usual receding horizon fashion.

Algorithm 2.5 (Model Predictive Control for Continuous-Time Systems). *At each sampling instant* $t_i = i\delta$, $i \in \mathbb{N}_0$, *measure the state* $x(t_i)$ *and solve Problem 2.4. Apply the input*

$$u_{\text{MPC}}(t) = u_T^*(t; x(t_i), t_i), \quad t_i \leq t < t_i + \delta. \tag{2.3}$$

to the nonlinear system (2.1) *until the next sampling instant* $t_{i+1} = t_i + \delta$.

In order to guarantee stability we require the following assumptions on our design parameters, i.e., the stage cost F, the terminal cost E, and the terminal region Ω.

Assumption 2.4. *The stage cost* $F : \mathbb{R}^n \times \mathbb{U} \to \mathbb{R}_{\geq 0}$ *is continuous,* $F(0,0) = 0$, *and there exists a class* $\mathcal{K}_\infty$ *function* $\underline{\alpha}_F : \mathbb{R}_{\geq 0} \to \mathbb{R}_{\geq 0}$ *such that* $F(x,u) \geq \underline{\alpha}_F(|x|)$ *for all* $x \in \mathbb{R}^n$, $u \in \mathbb{U}$. *The terminal region* Ω *is a closed set and contains* $0 \in \mathbb{R}^n$ *in its interior. The terminal cost function* $E : \mathbb{R}^n \to \mathbb{R}_{\geq 0}$ *is continuously differentiable and positive definite.*

Assumption 2.5. *For the nonlinear system* (2.1), *there exists a locally asymptotically stabilizing controller* $u = k(x)$ *such that*

a) the terminal region Ω *is controlled positively invariant under the control* $u = k(x)$,

b) $k(x) \in \mathbb{U}$ *for all* $x \in \Omega$, *and*

c) the extended Lyapunov inequality $\dot{E}(x) \leq -F(x, k(x))$ *is satisfied for all* $x \in \Omega$.

If the design conditions above are satisfied, asymptotic stability of the closed-loop can be established as stated in the following theorem.

Theorem 2.6 (Stability of MPC). *Consider the nonlinear system* (2.1) *and suppose that Assumptions 2.1–2.5 are satisfied. Then, the closed-loop system resulting from the application of the model predictive controller according to Algorithm 2.5 to system* (2.1) *is asymptotically stable. The region of attraction is the set of all initial conditions for which Problem 2.4 is initially feasible at* $t_i = 0$.

The proof can be found, e.g., in (Chen, 1997; Chen and Allgöwer, 1998; Findeisen et al., 2003; Fontes, 2001), see also the discrete-time versions in (Grüne and Pannek, 2011; Mayne et al., 2000; Rawlings and Mayne, 2009).

The first goal of this thesis is to provide alternative MPC formulations with guaranteed stability which complement the framework presented in this section, see Chapter 3. The second goal is to extend the available stability results to the class of nonlinear time-delay systems and, in particular, overcome the difficulties introduced due to their infinite-dimensional nature, see Chapter 4. In the subsequent section, we provide the necessary background of time-delay systems and associated stability conditions.

2.2 Time-Delay Systems

Nonlinear time-delay systems, also commonly referred to as differential-difference equations or systems with aftereffect or dead-time (Richard, 2003), naturally arise in the modelling of many technical, biological and social systems, for which the future evolution of the states does not only depend on the current state, but also on its past history. Technical systems with time-delays appear, e.g., when communication and computational delays are present in control loops or when transportation of material is considered such as in a continuous stirred tank reactor with recycle stream. Examples in the area of physics include polymer crystallization, laser models, and relativistic dynamics. Other examples from biology and social sciences are disease transmission models, glucose-insulin regulation, population dynamics, and price fluctuations. More details on the aforementioned examples and a comprehensive list of other examples can be found in (Kolmanovskii and Myshkis, 1999, Chapter 2).

In Section 2.2.1, we introduce retarded functional differential equations as the mathematical tool to describe continuous-time nonlinear time-delay systems considered in this thesis. The stability analysis of such systems is considered in Section 2.2.2, while Section 2.2.3 provides a brief overview of corresponding control methods available in the literature. For more details, we refer to the books (Gu et al., 2003; Hale, 1977; Hale and Lunel, 1993; Kolmanovskii and Myshkis, 1999).

2.2.1 Retarded Functional Differential Equations

Given $\tau \in \mathbb{R}_{>0}$, let $\mathcal{C}_\tau = \mathcal{C}([-\tau, 0], \mathbb{R}^n)$ denote the Banach space of continuous functions mapping the interval $[-\tau, 0] \subset \mathbb{R}$ into $\mathbb{R}^n$. A segment $x_t \in \mathcal{C}_\tau$ is defined by $x_t(\theta) = x(t + \theta), \theta \in [-\tau, 0]$. The norm on $\mathcal{C}_\tau$ is defined as $\|x_t\|_\tau = \sup_{\theta \in [-\tau, 0]} |x(t + \theta)|$. An autonomous *retarded functional differential equation* (RFDE) with initial function $\varphi \in \mathcal{C}_\tau$ can then be written in the general form

$$\dot{x}(t) = f(x_t), \tag{2.4a}$$
$$x(\theta) = \varphi(\theta), \qquad \forall \theta \in [-\tau, 0]. \tag{2.4b}$$

Note that these systems are infinite-dimensional in contrast to the finite-dimensional system (2.1) described by an ODE. As an important special case, we particularly consider systems with one constant discrete delay, which can be described by the RFDE

$$\dot{x} = f(x(t), x(t - \tau))$$

with time-delay $\tau \in \mathbb{R}_{>0}$. There are several results available regarding the existence and uniqueness of solutions of system (2.4), mostly relying on Lipschitz continuity assumptions on f. For more details, we refer to (Kolmanovskii and Myshkis, 1999, Chapter 3).

2.2.2 Stability

Stability and asymptotic stability of an RFDE is defined using an ε-δ condition analogously to the finite-dimensional case, see Definition 2.2 and Khalil (2002); Vidyasagar (1993), with the finite-dimensional norm replaced by $\|\cdot\|_\tau$ in the definition (Kolmanovskii and Myshkis, 1999). Without loss of generality, we can assume that the equilibrium of system (2.4) is given by $x_{t,s} = 0 \in \mathcal{C}_\tau$, i.e., $f(0) = 0$.

Definition 2.7 (Stability). *The equilibrium $x_{t,s} = 0$ of system (2.4) is called* stable *if for any $\varepsilon \in \mathbb{R}_{>0}$ there exists a $\delta_\varepsilon \in \mathbb{R}_{>0}$ such that $\|\varphi\|_\tau \leq \delta_\varepsilon$ implies that $|x(t)| \leq \varepsilon$ for all $t \geq 0$. It is called* asymptotically stable *if it is stable and there exists a $\delta_0 \in \mathbb{R}_{>0}$ such that $\|\varphi\|_\tau \leq \delta_0$ implies that $\lim_{t \to \infty} |x(t)| = 0$.*

Remark 2.8. *As in the finite-dimensional case, we sometimes refer to (asymptotic) stability of a system meaning (asymptotic) stability of the equilibrium at the origin.*

There are two types of Lyapunov theorems for time-delay systems, namely theorems based on Lyapunov-Krasovskii functionals and theorems using Lyapunov-Razumikhin functions.

Lyapunov-Krasovskii theory is the natural extension of Lyapunov's theorem towards systems with time-delays and is based on non-increasing Lyapunov-Krasovskii-functionals. In contrast, Lyapunov-Razumikhin uses continuous functions instead of functionals. For the sake of self-containedness, we repeat both theorems in the following. The interested reader is referred to Gu et al. (2003); Hale and Lunel (1993) and Kolmanovskii and Myshkis (1999) for further details and the corresponding proofs.

In order to state the stability theorems, we require the following definitions. The upper right-hand Dini derivative of the functional $E : \mathcal{C}_\tau \to \mathbb{R}_{\geq 0}$ along the solutions of (2.4) is defined by

$$\dot{E}(x_t) = \limsup_{\delta \to 0^+} \frac{E(x_{t+\delta}) - E(x_t)}{\delta} .$$

The upper right-hand Dini derivative of the function $V : \mathbb{R}^n \to \mathbb{R}_{\geq 0}$ with respect to (2.4) is defined by

$$\dot{V}(x_t) = \limsup_{\delta \to 0^+} \frac{V(x(t) + \delta f(x_t)) - V(x(t))}{\delta} .$$

With these definitions, we can repeat the following well-known results (Gu et al., 2003; Hale and Lunel, 1993; Kolmanovskii and Myshkis, 1999).

Theorem 2.9 (Lyapunov-Krasovskii). *Suppose that f maps bounded sets $\mathfrak{C} \subset \mathcal{C}_\tau$ into bounded sets in $\mathbb{R}^n$, and $\alpha_1, \alpha_2, \alpha_3 : \mathbb{R}_{\geq 0} \to \mathbb{R}_{\geq 0}$ are continuous, non-decreasing functions with $\alpha_1(0) = \alpha_2(0) = \alpha_3(0) = 0$ and $\alpha_1(s) > 0, \alpha_2(s) > 0$ for $s > 0$. If there exists a continuous functional $E : \mathcal{C}_\tau \to \mathbb{R}_{\geq 0}$ such that*

$$\alpha_1(|x(t)|) \leq E(x_t) \leq \alpha_2(\|x_t\|_\tau) ,$$
$$\dot{E}(x_t) \leq -\alpha_3(|x(t)|) ,$$

then the equilibrium $x_{t,s} = 0$ of (2.4) is stable. If, in addition, $\alpha_3(s) > 0$ for $s > 0$, then it is asymptotically stable. Furthermore, if additionally $\alpha_1(s) \to \infty$ as $s \to \infty$, then it is globally asymptotically stable.

Theorem 2.10 (Lyapunov-Razumikhin). *Suppose that f maps bounded sets $\mathfrak{C} \subset \mathcal{C}_\tau$ into bounded sets in $\mathbb{R}^n$, and $\alpha_1, \alpha_2 : \mathbb{R}_{\geq 0} \to \mathbb{R}_{\geq 0}$ are continuous, non-decreasing functions with $\alpha_1(0) = \alpha_2(0) = 0$ and $\alpha_1(s) > 0$ for $s > 0$. If there exists a continuous function $V : \mathbb{R}^n \to \mathbb{R}_{\geq 0}$ such that*

$$\alpha_1(|x(t)|) \leq V(x(t)) ,$$
$$\dot{V}(x_t) \leq -\alpha_2(|x(t)|) \quad \text{whenever } \forall \theta \in [-\tau, 0] : V(x(t + \theta)) \leq V(x(t)) ,$$

then the equilibrium $x_{t,s} = 0$ of (2.4) is stable. If, in addition, $\alpha_2(s) > 0$ for $s > 0$ and there exists $\rho \in \mathbb{R}_{>1}$ such that

$$\dot{V}(x_t) \leq -\alpha_2(|x(t)|) \quad \text{whenever } \forall \theta \in [-\tau, 0] : V(x(t + \theta)) \leq \rho\, V(x(t)) ,$$

then it is asymptotically stable. Furthermore, if additionally $\alpha_1(s) \to \infty$ as $s \to \infty$, then it is globally asymptotically stable.

2.2.3 Control of Nonlinear Time-Delay Systems

Due to the infinite-dimensional nature of time-delay systems, even the stability analysis of linear time-delay systems is already a challenging problem. Consequently, for the control of nonlinear time-delay systems only methods considering special cases can be expected. Several methods known from nonlinear control theory have been extended towards time-delay systems, including feedback linearization (Germani et al., 2000; Márquez-Martínez and Moog, 2004; Oguchi et al., 2002), backstepping (Mazenc and Bliman, 2006), flatness based control (Küchler and Sawodny, 2010; Rudolph, 2005; Rudolph and Winkler, 2003), control Lyapunov function approaches (Hua et al., 2008; Jankovic, 2001, 2003, 2005), and sum of squares techniques (Papachristodoulou, 2004, 2005). In all of the aforementioned control methods, it is not trivial to guarantee satisfaction of hard input and/or state constraints. For problems with such requirements, MPC is an attractive and natural choice. In that respect, it is interesting to note that there exists a significant amount of literature considering MPC for linear time-delay systems, e.g., Han et al. (2008); Hu and Chen (2004); Jeong and Park (2005); Kwon et al. (2003, 2004); Lee et al. (2011); Li and Xi (2011); Mahboobi Esfanjani and Nikravesh (2009b, 2010); Shi et al. (2009); Zhilin et al. (2003), to mention only a few. On the contrary, only a significantly smaller number of publications is available in the area of MPC for nonlinear time-delay systems. Besides our previous publications summarized in Chapter 4, this problem has only been considered in (Angrick, 2007; Kwon et al., 2001a,b; Lu, 2011; Mahboobi Esfanjani and Nikravesh, 2009a, 2011; Raff et al., 2007).

2.3 Summary

In this chapter, we have recalled the background necessary for the remaining chapters of this thesis.

First, we have started in Section 2.1 by introducing the basic idea of MPC, giving a brief literature overview of MPC schemes, and recalling the currently most well-established framework for MPC with guaranteed nominal stability. Alternatives to this framework are presented in Chapter 3, in which we propose five novel MPC schemes with guaranteed nominal stability for nonlinear continuous-time systems. We begin with the simplest MPC setup, i.e., unconstrained MPC without terminal cost and without terminal constraints. We derive explicit bounds for a minimal stabilizing prediction horizon for continuous-time systems and thereby extend existing results for discrete-time systems. Due to the conservative nature of these bounds, we investigate the use of additional weighting terms in order to obtain better stability conditions. Furthermore, we are interested in connections between these MPC schemes based on a controllability assumption and the classical MPC setup with terminal cost terms and terminal constraints, which have so far been treated separately in the literature.

Second, we have discussed time-delay systems in Section 2.2 with a particular focus on stability and control. Since there are only very few results on the control of nonlinear time-delay systems with hard constraints, we investigate the use of MPC for this class of systems in Chapter 4.

Chapter 3

Model Predictive Control for Nonlinear Continuous-Time Systems

In this chapter, we consider model predictive control for nonlinear continuous-time systems which are described by first-order ordinary differential equations. When considering the currently available MPC literature as reviewed in Section 2.1.2, one can distinguish two main existing classes of MPC schemes with guaranteed stability, see Figure 3.1. The first class of schemes uses a combination of *terminal cost* terms and *terminal constraints*. Roughly speaking, a control Lyapunov function (CLF), or functional in the case of infinite-dimensional systems, employed as additional terminal cost term ensures the cost of the finite horizon optimal control problem to be an upper bound on the infinite horizon cost. In most cases, the control Lyapunov function is only defined in some terminal set around the origin and, hence, additional terminal constraints are added to the optimization problem. For more details, we refer to Section 2.1.3 and the references therein. The second class of schemes uses a *controllability assumption* in terms of the stage cost instead. This controllability assumption is a less restrictive assumption compared to the use of a control Lyapunov function. However, in the existing literature the controllability assumption has to be valid at least in an invariant region containing the initial condition. Explicit conditions on the length of the prediction horizon for guaranteed stability can be derived, e.g., by solving a so-called *abstract* linear program. For more details, see Section 3.2 and the references therein. In the following, we refer to the first class of schemes as *CLF-MPC* and we use the notion of *unconstrained MPC* for the second class as common in the literature (due to the lack of stabilizing terminal constraints and regardless of the presence of input and/or state constraints).

While CLF-MPC can be considered well-understood for both discrete-time and continuous-time systems, for unconstrained MPC explicit bounds on a stabilizing prediction horizon are only available for discrete-time systems. For continuous-time systems, the abstract linear program becomes infinite-dimensional, but can nevertheless be solved due to its particular structure as will be proven in Section 3.2. We also show that several results for discrete-time systems, such as suboptimality estimates and benefits of a growth condition, can be extended to continuous-time systems. Furthermore, we are interested in connections between CLF-MPC and unconstrained MPC. In the literature, these two classes have almost exclusively been considered separately. In Sections 3.3 and 3.4, we show that additional weighting terms in the MPC setup can yield improved stability conditions for unconstrained MPC. In Sections 3.5 and 3.6, we investigate the use of only local controllability assumptions in combination with terminal constraints and additional weighting terms in order to guarantee stability. See Figure 3.1 for a schematic overview of the different MPC schemes with guaranteed stability.

The remainder of this chapter is organized as follows. The problem setup and necessary assumptions on the system are described in detail in Section 3.1. In the subsequent sections, we introduce five different MPC setups and provide novel conditions for each setup, which guarantee asymptotic stability of the closed-loop. In Section 3.2, we consider the simplest MPC setup without terminal cost and without terminal constraints. This setup is particularly interesting due to its simplicity, the frequent use in practical applications, and the simple calculation of performance estimates of the closed-loop. In Section 3.3, stability conditions for unconstrained MPC with a positive semi-definite terminal cost function are derived. We show that if the terminal cost is in a particular sense "similar" to a control Lyapunov function, stability can be guaranteed for shorter prediction horizons. For the same goal, Section 3.4 introduces an exponential weighting on the stage cost for improved stability conditions. In Sections 3.5 and 3.6, additional weighting terms are presented which guarantee stability with only local controllability assumptions in combination with terminal constraints. Furthermore, the integral terminal cost term considered in Section 3.5 allows to consider both previous classes of MPC schemes (CLF-MPC and unconstrained MPC) in a unified way. Finally, we summarize the results of this chapter in Section 3.7.

Parts of this chapter are based on Reble and Allgöwer (2011, 2012b); Reble et al. (2012a,b).

3.1 Problem Setup

In this chapter, we consider nonlinear continuous-time systems described by the ordinary differential equation (ODE)

$$\dot{x}(t) = f(x(t), u(t)), \tag{3.1a}$$

$$x(0) = x_0, \tag{3.1b}$$

in which $x(t) \in \mathbb{R}^n$ is the state at time t, $x_0 \in \mathbb{R}^n$ is the initial condition, and $u(t) \in \mathbb{R}^m$ is the control input subject to input constraints $u(t) \in \mathbb{U} \subset \mathbb{R}^m$.

We will use the following three standard assumptions throughout the remainder of this chapter.

Assumption 3.1. *The vector field* $f : \mathbb{R}^n \times \mathbb{R}^m \to \mathbb{R}^n$ *is continuously differentiable and* $f(0,0) = 0$, *i.e.,* $x_s = 0$ *is an equilibrium of system (3.1) for* $u_s = 0$.

Assumption 3.2. *System (3.1) has a unique solution for any initial condition* $x_0 \in \mathbb{R}^n$ *and any piecewise- and right-continuous input function* $u : \mathbb{R}_{\geq 0} \to \mathbb{U}$.

Assumption 3.3. *The input constraint set* $\mathbb{U} \subset \mathbb{R}^m$ *is compact and contains the origin.*

The problem of interest is to stabilize the steady state $x_s = 0$ and to achieve some optimal performance via model predictive control. A classical, well-established result on MPC for nonlinear systems with guaranteed stability was recalled in Theorem 2.6. In this chapter, we provide alternative MPC formulations which guarantee stability of the closed-loop and discuss advantages and disadvantages of these schemes.

Remark 3.1. *In Assumption 3.1, we require the origin to be a steady state of the system, i.e.,* $f(0,0) = 0$. *This is without loss of generality because for any other steady state* $(x_s, u_s) \in \mathbb{R}^n \times \mathbb{R}^m$ *of the system, i.e.,* $f(x_s, u_s) = 0$, *this case can be recovered by using the simple coordinate transformation* $\tilde{x} = x - x_s$, $\tilde{u} = u - u_s$, *see (Khalil, 2002, Section 4.1) for more details.*

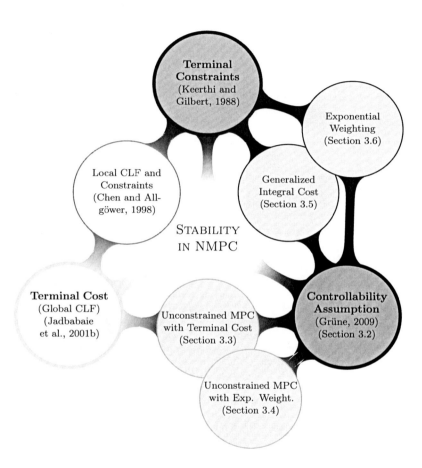

Figure 3.1: Schematic overview of stability conditions in MPC.

Remark 3.2. *For the sake of a concise presentation, we restrict ourselves to finite-dimensional continuous-time systems in this chapter although the results also hold – mutatis mutandis – if the state and input are elements of arbitrary metric spaces, in particular also infinite-dimensional spaces. The main change necessary is to replace the norm $|\cdot|$ on $\mathbb{R}^n$ by an appropriate norm of the corresponding space. However, in some cases other setups are desirable. In particular in MPC for nonlinear time-delay systems, as considered in Chapter 4, several assumptions cannot be satisfied in general, see also Remark 4.28 in Section 4.5.*

Remark 3.3. *In the MPC literature it is often assumed that the input constraint set $\mathbb{U}$ contains the origin in its interior. For instance, this is necessary when calculating a terminal cost function based on the Jacobi linearization of the system (Chen and Allgöwer, 1998). However, it is not required for the results in this chapter.*

3.2 Unconstrained MPC and Suboptimality Estimates

In this section, we consider the simplest MPC setup for nonlinear systems, namely, MPC without terminal cost terms and terminal constraints, which is commonly referred to as *unconstrained MPC* (Grüne, 2009; Grüne et al., 2010a). This scheme is particularly crucial due to its widespread use in practical applications. It is well known that the closed-loop is asymptotically stable under mild assumptions if the prediction horizon is chosen large enough (Jadbabaie and Hauser, 2005). However, no explicit bounds on a minimal stabilizing prediction horizon have been given for continuous-time systems in the literature. We extend the results for discrete-time systems presented in (Grüne, 2009; Grüne et al., 2010a) and derive explicit conditions on the prediction horizon based on a controllability assumption of the system and two corresponding infinite-dimensional linear programs. Here, the infinite-dimensionality represents the main difficulty compared to the discrete-time setup. However, the particular structures of both linear programs allow solutions involving only a single integration of a scalar variable related to the controllability assumption. Furthermore, the unconstrained MPC setup allows to give guaranteed bounds on the performance compared to an infinite horizon optimal control law. For the special case of an exponential controllability assumption, we can explicitly solve the integrals and give analytic expressions of the suboptimality estimate in terms of the prediction horizon and the sampling time. Finally, we compare our results to the results of Grüne et al. (2010b,c), in which the methods for discrete-time systems are applied to continuous-time systems in a sampled-data context. A more thorough discussion of connections between the results in this section and the previous results in discrete-time (Grüne, 2009; Grüne et al., 2010a) can be found in (Worthmann, Reble, Grüne, and Allgöwer, 2012).

3.2.1 Unconstrained MPC Setup

One main advantage of unconstrained MPC schemes without terminal cost is the possibility to directly conclude performance guarantees in terms of a suboptimality estimate. With respect to performance, we want to minimize the infinite horizon cost functional $J_\infty(x_0, u)$

$$\underset{u \in \mathcal{PC}(\mathbb{R}_{\geq 0}, \mathbb{U})}{\text{minimize}} J_\infty(x_0, u), \qquad \text{in which } J_\infty(x_0, u) = \int_0^\infty F(x(t), u(t)) \, dt,$$

subject to the system dynamics (3.1). The associated optimal cost of this infinite horizon optimal control problem is denoted by $J_\infty^\star(x_0) = \underset{u \in \mathcal{PC}(\mathbb{R}_{\geq 0}, \mathbb{U})}{\min} J_\infty(x_0, u)$ in this section and we use the following standard condition concerning the stage cost F.

Assumption 3.4. *The stage cost $F : \mathbb{R}^n \times \mathbb{U} \to \mathbb{R}_{\geq 0}$ is continuous, $F(0, 0) = 0$, and there exists a class $\mathcal{K}_\infty$ function $\underline{\alpha}_F : \mathbb{R}_{\geq 0} \to \mathbb{R}_{\geq 0}$ such that*

$$F(x, u) \geq F(x, 0) \geq \underline{\alpha}_F(|x|) \quad \text{for all } x \in \mathbb{R}^n, \, u \in \mathbb{U}. \tag{3.2}$$

Remark 3.4. *Assumption 3.4 can be relaxed to a positive semi-definite stage cost F in combination with a suitable detectability condition (Grimm et al., 2005). For a concise presentation, we use the stricter assumption of a positive definite stage cost.*

Since infinite horizon problems are often computationally intractable, finite horizon cost functionals combined with a receding horizon strategy are often used instead. The finite horizon cost functional used in this section is given by

$$J_T(x_0, u) = \int_0^T F(x(t), u(t)) \, dt,$$

in which T is the prediction horizon. The open-loop finite horizon optimal control problem at sampling instant t_i given the measured state $x(t_i)$ is now formulated as follows.

Problem 3.5.

$$\underset{\bar{u} \in \mathcal{PC}([t_i, t_i + T], \mathbb{R}^m)}{\text{minimize}} J_T(x(t_i), \bar{u}) \tag{3.3a}$$

subject to

$$\dot{\bar{x}}(t'; x(t_i), t_i) = f(\bar{x}(t'; x(t_i), t_i), \bar{u}(t')), \qquad t' \in [t_i, t_i + T], \tag{3.3b}$$

$$\bar{x}(t_i; x(t_i), t_i) = x(t_i), \tag{3.3c}$$

$$\bar{u}(t') \in \mathbb{U}, \qquad t' \in [t_i, t_i + T], \tag{3.3d}$$

in which

$$J_T(x(t_i), \bar{u}) = \int_{t_i}^{t_i + T} F(\bar{x}(t'; x(t_i), t_i), \bar{u}(t')) \, dt'. \tag{3.3e}$$

In Problem 3.5, $\bar{x}(t'; x(t_i), t_i)$ denotes the predicted trajectory starting from initial condition $\bar{x}(t_i; x(t_i), t_i) = x(t_i)$ and driven by $\bar{u}(t')$ for $t' \in [t_i, t_i + T]$. We assume that the optimal open-loop control which minimizes $J_T(x(t_i), \bar{u})$ is given by $u_T^\star(t'; x(t_i), t_i)$ for all $t' \in [t_i, t_i + T]$. The associated optimal cost is denoted by $J_T^\star(x(t_i))$ and the associated predicted trajectory is $x^\star(t'; x(t_i), t_i)$, $t' \in [t_i, t_i + T]$. For given sampling time δ with $0 < \delta \leq T$, the control input to the system is defined by the following algorithm in the usual receding horizon fashion.

Algorithm 3.6 (Unconstrained Model Predictive Control). *At each sampling instant $t_i = i\delta$, $i \in \mathbb{N}_0$, measure the state $x(t_i)$ and solve Problem 3.5. Apply the input*

$$u_{\mathrm{MPC}}(t) = u_T^*(t; x(t_i), t_i), \quad t_i \le t < t_i + \delta \tag{3.4}$$

to the system until the next sampling instant $t_{i+1} = t_i + \delta$.

Remark 3.7. *For the sake of simplicity, we consider a constant sampling time δ throughout the thesis. In contrast to the classical MPC setup, see Theorem 2.6, the conditions for convergence and asymptotic stability derived in this chapter depend explicitly on the sampling time. However, analogue results can be obtained for time-varying sampling times as in Grüne et al. (2010a), in which it is referred to as "time-varying control horizon".*

It is well known that such a definition of the control law is not guaranteed to yield an asymptotically stable closed-loop in general. A practical example for this fact is given in Raff et al. (2006). Besides guaranteed nominal stability, we are also interested in an evaluation of the performance of the resulting MPC controller. To this end, let

$$J_\infty^{\mathrm{MPC}}(x_0) = J_\infty(x_0, u_{\mathrm{MPC}})$$

denote the infinite horizon cost resulting from application of the MPC control law defined by Algorithm 3.6 to the nonlinear system (3.1). We define the suboptimality estimate α of the closed-loop as in Grüne (2009); Grüne et al. (2010a).

Definition 3.8 (Suboptimality Estimate). *If for a constant $\alpha \in \mathbb{R}$ and for all $x_0 \in \mathbb{R}^n$*

$$\alpha J_\infty^{\mathrm{MPC}}(x_0) \le J_\infty^*(x_0),$$

then we call α a suboptimality estimate *of the closed-loop.*

From this definition, it is clear that $\alpha \le 1$ because $J_\infty^{\mathrm{MPC}}(x_0) \ge J_\infty^*(x_0)$. Moreover, $\alpha = 1$ corresponds to infinite horizon optimality and, if $\alpha > 0$, then stability of the closed-loop is guaranteed as will be shown later. In the following section, we derive stability conditions and a suboptimality estimate for the closed-loop using the unconstrained MPC scheme presented in this section.

3.2.2 Asymptotic Stability and Suboptimality Estimate

In order to derive the suboptimality estimate α and to give stability conditions, we will use a result from relaxed dynamic programming and a suitable controllability assumption, which gives an upper bound on the optimal cost in terms of the stage cost. With this assumption, we can derive several properties of the optimal trajectories and, in particular, compare the optimal cost at two consecutive sampling instants. Loosely speaking, a decrease of the optimal cost guarantees stability and the desired suboptimality estimate.

In the subsequent analysis, we consider the two consecutive sampling instants $t_0 = 0$ and $t_1 = \delta$ without loss of generality. Since system (3.1) is time-invariant, all results hold analogously for any other two consecutive sampling instants t_i and t_{i+1}. Furthermore, we use the following abbreviation

$$F^*(t; t_i) = F(x^*(t; x(t_i), t_i), u_T^*(t; x(t_i), t_i)) \tag{3.5}$$

for $t_i \in \{0, \delta\}$ and for all $t \in [t_i, t_i + T]$. Note that from this definition, it directly follows that $J_T^*(x(t_i)) = \int_{t_i}^{t_i+T} F^*(t'; t_i) \, dt'$. In order to derive the suboptimality estimate α, we first recall a crucial result from relaxed dynamic programming, see Grüne and Rantzer (2008); Lincoln and Rantzer (2006).

Proposition 3.9 (Relaxed Dynamic Programming). *Suppose that*

$$J_T^*(x^*(\delta; x_0, 0)) - J_T^*(x_0) \leq -\alpha \int_0^\delta F(x^*(t'; x_0, 0), u_{\mathrm{MPC}}(t')) \, dt', \qquad (3.6)$$

holds for all $x_0 \in \mathbb{R}^n$ and for some constant $\alpha \in [0, 1]$, then the estimates

$$\alpha \, J_\infty^*(x_0) \leq \alpha \, J_\infty^{\mathrm{MPC}}(x_0) \leq J_T^*(x_0) \leq J_\infty^*(x_0)$$

hold for all $x_0 \in \mathbb{R}^n$.

Proof. The result is a slightly modified version of (Grüne and Rantzer, 2008, Proposition 2.2) and (Grüne, 2009, Proposition 2.4). The first and third inequalities follow directly from optimality of $J_\infty^*(x_0)$ and $J_T^*(x_0)$, respectively. The second inequality is obtained by invoking (3.6) for states along the trajectory of the closed-loop at the sampling instants, i.e., $x(t_i) = x(i\delta)$ for $i \in \mathbb{N}_0$ and by summing up from $i = 0$ to $i = \infty$. Here, we can exploit the telescoping series property since the series $J_T^*(x(i\delta))$ is convergent because it is non-increasing and bounded from below. $\square$

In order to derive α such that (3.6) holds, we use a controllability assumption of the system in terms of the stage cost similar to the controllability assumption used in discrete-time by Grüne (2009); Grüne et al. (2010a).

Assumption 3.5 (Controllability Assumption). *For all $T' \in \mathbb{R}_{\geq 0}$ and $x_0 \in \mathbb{R}^n$, there exists a piece-wise continuous input trajectory $\hat{u}(\cdot; x_0, 0)$ with $\hat{u}(t; x_0, 0) \in \mathbb{U}$ for all $t \in [0, T']$ and*

$$J_{T'}^*(x_0) \leq J_{T'}(x_0, \hat{u}) \leq B(T')F(x_0, 0), \qquad (3.7)$$

in which $B : \mathbb{R}_{\geq 0} \to \mathbb{R}_{>0}$ is a continuous, non-decreasing, and bounded function.

Assumption 3.5 is directly related to the assumption used in (Grimm et al., 2005, Corollary 3) for discrete-time systems and is slightly more general than the controllability assumption used in (Grüne, 2009; Grüne et al., 2010a), in which an upper bound on the stage cost was considered in contrast to an upper bound on the optimal cost $J_{T'}^*$. The results based on this more general assumption allow to directly obtain better suboptimality estimates when using a growth condition in addition to the assumption of exponential controllability, see also Assumption 3.6 in Section 3.2.3 and the following discussion. Furthermore, it has been shown by Worthmann (2012a) that this more general assumption is indeed also sufficient for the discrete-time results obtained by (Grüne, 2009; Grüne et al., 2010a) despite being less restrictive.

Remark 3.10. *We do not consider state constraints throughout this chapter and use "global" controllability assumptions for all $x \in \mathbb{R}^n$ for a concise presentation in this section as well as Sections 3.3 and 3.4. Modifications and "regional" versions using invariant sets containing the initial state x_0 can be obtained in a straightforward manner. For semiglobal stability results in a discrete-time setting, see, e.g., (Grüne and Pannek, 2011, Section 6.7).*

Remark 3.11. *There exist several possibilities to verify the controllability assumption. Several examples are considered in Grüne and Pannek (2011) and the application of unconstrained MPC to a fixed-wing UAV was investigated in Halter (2012). Other approaches are given by suboptimal explicit off-line controller design methods (Johansen, 2004) or interval arithmetic methods (Schulze Darup and Mönnigmann, 2011; Summers et al., 2010).*

The following lemma gives a direct consequence of Controllability Assumption 3.5.

Lemma 3.12 (Implications of Controllability Assumption 3.5). *Suppose that Assumptions 3.1–3.4 and Controllability Assumption 3.5 are satisfied. Then,*

$$J_T^*(x(\delta)) \leq \int_\delta^{t^*} F^*(t';0)dt' + B(T + \delta - t^*)F^*(t^*;0) \tag{3.8a}$$

holds for all $t^* \in [\delta, T]$ and

$$\int_{t^*}^{T} F^*(t';0)dt' \leq B(T - t^*)\, F^*(t^*;0) \tag{3.8b}$$

holds for all $t^* \in [0, T]$.

Proof. For any $t^* \in [\delta, T]$ define the control trajectory $\tilde{u}_{t^*}$ by

$$\tilde{u}_{t^*}(t) = \begin{cases} u_T^*(t; x(0), 0), & t \in [\delta, t^*[\\ \widehat{u}(t - t^*; x^*(t^*; x(0), 0)), & t \in [t^*, T + \delta[\end{cases}$$

in which $\widehat{u}$ is the input trajectory from Assumption 3.5 for initial state $x^*(t^*; x(0), 0)$. Since $\tilde{u}_{t^*}$ is a feasible, but not necessarily optimal, solution to the finite horizon optimal control problem 3.5 for initial state $x(\delta)$, we obtain

$$J_T^*(x(\delta)) \leq J_T(x(\delta), \tilde{u}_{t^*}) \leq \int_\delta^{t^*} F^*(t';0)dt' + J_{T+\delta-t^*}(x^*(t^*; x(0), 0), \widehat{u})$$

$$\overset{(3.7)}{\leq} \int_\delta^{t^*} F^*(t';0)dt' + B(T + \delta - t^*)F^*(t^*;0)\,,$$

which proves (3.8a). The proof of (3.8b) follows directly from optimality of $J_T^*(x(0))$, the principle of optimality (endpieces of optimal trajectories are optimal), see (Bellman, 1957, Chapter III, §3), and Controllability Assumption 3.5. For a similar result, see, e.g., (Reble and Allgöwer, 2011, Lemma 3). □

In order to obtain a suboptimality estimate α, we will use the following fact.

Proposition 3.13. *Suppose that Assumptions 3.1–3.4 and Controllability Assumption 3.5 are satisfied. If α is a lower bound (or ideally equal) to the minimum value of the infinite-dimensional linear program*

$$\min_{F^*(\cdot;0), J_T^*(x(\delta))} \frac{\int_0^T F^*(t';0)dt' - J_T^*(x(\delta))}{\int_0^\delta F^*(t';0)dt'} \quad \text{subject to (3.8)}\,,\ J_T^*(x(\delta)) > 0\,,$$

then (3.6) in Proposition 3.9 holds.

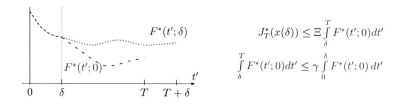

$$J_T^*(x(\delta)) \leq \Xi \int_\delta^T F^*(t';0)dt'$$

$$\int_\delta^T F^*(t';0)dt' \leq \gamma \int_0^\delta F^*(t';0)\,dt'$$

Figure 3.2: Sketch for Lemmata 3.14 and 3.15. The optimal trajectory calculated at time 0 is depicted by the (green and red) dashed line, the optimal trajectory calculated at δ is depicted by the (blue) dotted line.

Proof. The constraints $J_T^*(x(\delta)) > 0$ and (3.8) ensure that $F^*(t';0) > 0$ for all $t' \in [0,\delta]$. Hence, $\int_0^\delta F^*(t';0)dt' > 0$ and the result follows immediately. $\qquad\square$

The infinite-dimensional program corresponds to the linear program in (Grüne, 2009, Problem 4.4) concerning unconstrained MPC for discrete-time systems. We will derive a lower bound and conditions, for which this lower bound is indeed the minimum value. To this end, we now state two intermediate results in Lemmata 3.14 and 3.15 based upon the controllability assumption. Lemma 3.14 uses the optimality of $J_T^*(x(\delta))$ in addition to the controllability assumption in order to derive an upper bound on $J_T^*(x(\delta))$ in terms of the endpiece of the predicted trajectory calculated at time $t_0 = 0$. In Figure 3.2, this can be interpreted as giving an upper bound on the integral cost of the blue dotted line in terms of the red loosely dashed line. Lemma 3.15 applies the principle of optimality (Bellman, 1957, Chapter III, §3), which states that the trajectory $F^*(t;0)$ calculated at time $t_0 = 0$ is an optimal endpiece on the interval $[\delta, T]$. Hence, it is possible to derive a similar upper bound based on the controllability assumption. In Figure 3.2, the result can be interpreted as giving an upper bound on the cost of the red loosely dashed line in terms of the green dashed line.

Lemma 3.14 (Calculation of Ξ). *Suppose that Assumptions 3.1–3.4 and Controllability Assumption 3.5 are satisfied. Then,*

$$J_T^*(x(\delta)) = \int_\delta^{\delta+T} F^*(t';\delta)\,dt' \leq \Xi \int_\delta^T F^*(t';0)dt' \,, \tag{3.9}$$

in which

$$\frac{1}{\Xi} = 1 - \exp\left(-\int_\delta^T \frac{1}{B(T+\delta-t^*)}\,dt^*\right). \tag{3.10}$$

Proof. Due to Lemma 3.12, we know that (3.8a) holds. Consider any piece-wise continuous function $F^*(t;0) : [\delta, T] \to \mathbb{R}_{\geq 0}$ satisfying (3.8a) for a given arbitrary, but fixed, $J_T^*(x(\delta)) \geq 0$ and define the function $\widehat{F}^*(t) : [\delta, T] \to \mathbb{R}_{\geq 0}$ by

$$\widehat{F}^*(t) = \frac{J_T^*(x(\delta))}{B(T+\delta-t)}e^{-\int_\delta^t \frac{1}{B(T+\delta-t')}\,dt'}. \tag{3.11}$$

In the next step, we show that for all piece-wise continuous functions $F^*(t;0)$ satisfying (3.8a), the following holds

$$\int_\delta^T \widehat{F}^*(t')dt' \leq \int_\delta^T F^*(t';0)dt' \,. \tag{3.12}$$

To this end, note that $\widehat{F}^*(t)$ satisfies (3.8a) with equality (instead of inequality) for all $t^* \in [\delta, T]$, i.e.,

$$J_T^*(x(\delta)) = \int_\delta^{t^*} \widehat{F}^*(t')dt' + B(T + \delta - t^*)\widehat{F}^*(t^*) \,. \tag{3.13}$$

This can be shown by direct evaluation taking into account the anti-derivative of $\widehat{F}^*$

$$\int^t \widehat{F}^*(t')dt' = -J_T^*(x(\delta))\, e^{-\int_\delta^t \frac{1}{B(T+\delta-t')}dt'} + \mathfrak{C}$$

with $\mathfrak{C} \in \mathbb{R}$. For the sake of contradiction, assume now $\int_\delta^T \widehat{F}^*(t')dt' > \int_\delta^T F^*(t';0)dt'$. But then there exists a $t \in [\delta, T]$ for which

$$\int_\delta^t \widehat{F}^*(t')dt' \geq \int_\delta^t F^*(t';0)dt' \quad \text{and} \quad \widehat{F}^*(t) > F^*(t;0) \,,$$

and, consequently, due to (3.13) and $B(T + \delta - t) > 0$

$$J_T^*(x(\delta)) = \int_\delta^t \widehat{F}^*(t')dt' + B(T + \delta - t)\widehat{F}^*(t^*) > \int_\delta^t F^*(t';0)dt' + B(T + \delta - t)F^*(t;0).$$

But this contradicts (3.8a), which shows that (3.12) holds. On the other hand, direct calculations using the anti-derivative of $\widehat{F}^*$ reveal that

$$\int_\delta^T \widehat{F}^*(t')dt' \underset{\equiv}{=} \frac{1}{\equiv} J_T^*(x(\delta)) \,. \tag{3.14}$$

Combining (3.12) and (3.14) yields (3.9). This completes the proof. $\qquad\square$

Lemma 3.15 (Calculation of γ). *Suppose that Assumptions 3.1–3.4 and Controllability Assumption 3.5 are satisfied. Then,*

$$\int_\delta^T F^*(t';0)dt' \leq \gamma \int_0^\delta F^*(t';0)\,dt' \,, \tag{3.15}$$

in which

$$\frac{1}{\gamma} = \exp\left(\int_0^\delta \frac{1}{B(T - t^*)}\,dt^*\right) - 1 \,. \tag{3.16}$$

Proof. Due to Lemma 3.12, we know that (3.8b) holds. Consider any piece-wise continuous function $F^*(t; 0) : [0, T] \to \mathbb{R}_{\geq 0}$ satisfying (3.8b) for a given arbitrary, but fixed, $\int_{\delta}^{T} F^*(t'; 0) dt' \geq 0$ and define the function $\widehat{F}^*(t) : [0, \delta] \to \mathbb{R}_{\geq 0}$ by

$$\widehat{F}^*(t) = \frac{\int\limits_{\delta}^{T} F^*(t'; 0) dt'}{B(T - t)} e^{-\int\limits_{\delta}^{t} \frac{1}{B(T - t')} dt'}. \tag{3.17}$$

In the next step, we show that for all piece-wise continuous functions $F^*(t; 0)$ satisfying (3.8b), the following holds

$$\int\limits_{0}^{\delta} \widehat{F}^*(t') dt' \leq \int\limits_{0}^{\delta} F^*(t'; 0) dt'. \tag{3.18}$$

To this end, note that the anti-derivative of $\widehat{F}^*$ is

$$\int\limits^{t} \widehat{F}^*(t') dt' = -\int\limits_{\delta}^{T} F^*(t'; 0) dt' \, e^{-\int\limits_{\delta}^{t} \frac{1}{B(T - t')} dt'} + \mathfrak{C}$$

with $\mathfrak{C} \in \mathbb{R}$. Hence, $\widehat{F}^*$ satisfies (3.8b) with equality for all $t^* \in [0, \delta]$ in the sense of

$$\int\limits_{t^*}^{\delta} \widehat{F}^*(t') dt' + \int\limits_{\delta}^{T} F^*(t'; 0) dt' = B(T - t^*) \, \widehat{F}^*(t^*).$$

For any $F^*(t; 0)$ satisfying (3.8b), we know that $F^*(\delta; 0) - \widehat{F}^*(\delta) \geq 0$ and

$$\int\limits_{t^*}^{\delta} \left(F^*(t'; 0) - \widehat{F}^*(t') \right) dt' \leq B(T - t^*) \left(F^*(t^*; 0) - \widehat{F}^*(t^*) \right)$$

holds for all $t^* \in [0, \delta]$. Define $\mathfrak{F}(t^*) = F^*(\delta - t^*; 0) - \widehat{F}^*(\delta - t^*)$, for which $\mathfrak{F}(0) \geq 0$ and $\int_{0}^{t^*} \mathfrak{F}(t') dt' \leq B(T - \delta + t^*) \mathfrak{F}(t^*)$. Due to the comparison lemma (Khalil, 2002), it follows that $\mathfrak{F}(t^*) \geq 0$ for all $t^* \in [0, \delta]$. Consequently, Equation (3.18) holds. On the other hand, direct calculations using the anti-derivative of $\widehat{F}^*$ show that

$$\int\limits_{0}^{\delta} \widehat{F}^*(t') \, dt' = \frac{1}{\gamma} \int\limits_{\delta}^{T} F^*(t'; 0) dt', \tag{3.19}$$

with γ defined by (3.16). Hence, using (3.18) and (3.19) implies (3.15). This completes the proof. $\square$

The previous results allow us to state the main result of this section concerning the stability and suboptimality of the closed-loop.

Theorem 3.16 (Stability of Unconstrained MPC for Continuous-Time Systems). *Suppose that Assumptions 3.1–3.4 and Controllability Assumption 3.5 are satisfied for the nonlinear system* (3.1). *Furthermore, suppose that*

$$\alpha = 1 - \gamma \, (\Xi - 1) > 0 \, , \tag{3.20}$$

with Ξ and γ defined in Lemmata 3.14 and 3.15, respectively. Then, the closed-loop system resulting from the application of the model predictive controller according to Algorithm 3.6 to system (3.1) *is asymptotically stable and the suboptimality estimate*

$$\alpha \, J_\infty^{\mathrm{MPC}}(x_0) = \alpha \, J_\infty(x_0, u_{\mathrm{MPC}}) \leq J_\infty^*(x_0) \tag{3.21}$$

holds for all $x_0 \in \mathbb{R}^n$.

Proof. First, we note that $\Xi > 1$ in view of Equations (3.9) and (3.10). By using the results of Lemmata 3.14 and 3.15, respectively, we obtain

$$J_T^*(x(\delta)) - J_T^*(x(0)) = J_T^*(x(\delta)) - \int_0^T F^*(t'; 0) \, dt'$$

$$\stackrel{(3.9)}{\leq} (\Xi - 1) \int_\delta^T F^*(t'; 0) \, dt' - \int_0^\delta F^*(t'; 0) \, dt'$$

$$\stackrel{(3.15)}{\leq} \underbrace{(\gamma \, (\Xi - 1) - 1)}_{= -\alpha} \int_0^\delta F^*(t'; 0) dt'. \tag{3.22}$$

The suboptimality estimate (3.21) follows from Proposition 3.9. Asymptotic stability follows using similar arguments as in (Chen, 1997, Theorem 3.1). The optimal cost $J_T^*(x)$ is continuous in x at the origin, which can be shown analogously to (Chen, 1997, Lemma A.1). The solution of the open-loop system

$$\dot{\bar{x}}(t') = f(\bar{x}(t'), 0) \, , \quad x(0) = x_0$$

depends continuously on the initial condition x_0 (Khalil, 2002, Theorem 3.5) and $J_T^*(x)$ is non-increasing along trajectories of the closed-loop after the first sampling instant. This implies stability of the origin. Furthermore, stability implies existence of a $\delta_0 \in \mathbb{R}_{>0}$ such that $x(t)$ is bounded for all times if $|x_0| \leq \delta_0$. Since the input constraint set $\mathbb{U}$ is compact and f continuous, it follows that $f(x(t), u(t))$ is bounded for all $t \in \mathbb{R}_{>0}$. Hence, $x(t)$ and $\underline{\alpha}_F(|x(t)|)$ are uniformly continuous in t. On the other hand, $J_\infty^{\mathrm{MPC}}(x_0)$ is finite due to (3.21). Thus, the lower bound on the stage cost (3.2) implies

$$\int_0^\infty \underline{\alpha}_F(|x(t')|) dt' \leq J_\infty^{\mathrm{MPC}}(x_0) < \infty \, .$$

This implies $|x(t)| \to 0$ for $t \to \infty$ according to Barbalat's Lemma (Barbalat, 1959; Khalil, 2002). This completes the proof of asymptotic stability. $\square$

For α as in (3.20), $\alpha > 0$, and thereby asymptotic stability, can always be guaranteed for a prediction horizon chosen large enough, which is a well-known result, see, e.g., Jadbabaie and Hauser (2005). Moreover, we have $\Xi \to 1$ for $T \to \infty$ and, consequently, $\alpha \to 1$. This means that performance arbitrarily close to infinite horizon optimal performance can be achieved for a large enough (finite) prediction horizon. This observation is formally stated in the following proposition.

Proposition 3.17. *Suppose that Controllability Assumption 3.5 is satisfied. Then, there exists a $T^* \in \mathbb{R}_{>0}$ such that $\alpha > 0$ holds for all $T > T^*$ with α defined in Theorem 3.16. Furthermore, $\alpha \to 1$ for $T \to \infty$.*

Proof. Since $B(T)$ is bounded, i.e., there exists $B_\infty \in \mathbb{R}_{>0}$ such that $B(T) \leq B_\infty$ for all $T \in \mathbb{R}_{>0}$, we have for all $T \in \mathbb{R}_{>0}$

$$\gamma \leq \frac{1}{e^{\delta/B_\infty} - 1} = \gamma_\infty \quad \text{and} \quad \Xi - 1 \leq \frac{1}{e^{\frac{T-\delta}{B_\infty}} - 1}.$$

For $T > \delta + B_\infty \ln(\gamma_\infty + 1)$, it directly follows that $\alpha = 1 - \gamma(\Xi - 1) > 0$. Furthermore, for any $\varepsilon \in \mathbb{R}_{>0}$, we can define $T_\varepsilon = \delta + B_\infty \ln\left(\frac{\gamma_\infty}{\varepsilon} + 1\right)$, which guarantees $|\alpha - 1| < \varepsilon$ for all $T > T_\varepsilon$. This completes the proof. □

Our estimate α in Theorem 3.16 is the qualitatively best possible estimate, based on only the controllability assumption without any further information, in the following sense.

Theorem 3.18 (Connection of Suboptimality Estimate to Linear Program). *The suboptimality estimate α defined by (3.20) in Theorem 3.16 is the optimal value of the following infinite-dimensional program for any $J_T^*(x(\delta)) > 0$*

$$\min_{F^*(\cdot;0)} \frac{\int_0^T F^*(t';0)dt' - J_T^*(x(\delta))}{\int_0^\delta F^*(t';0)dt'} \tag{3.23}$$

subject to

$$J_T^*(x(\delta)) \leq \int_\delta^t F^*(i';0)dt' + B(T + \delta - t)F^*(t;0), \qquad \forall t \in [\delta, T], \tag{3.24a}$$

$$\int_t^T F^*(t';0)dt' \leq B(T - t)F^*(t;0), \qquad \forall t \in [0, \delta]. \tag{3.24b}$$

If, in addition, B satisfies

$$B(T + \delta - t)e^{-\int_t^T \frac{1}{B(T+\delta-t')}dt'} \geq B(T - t), \tag{3.25}$$

for all $t \in [\delta, T]$, then the same holds true when replacing $t \in [0, \delta]$ by $t \in [0, T]$ in (3.24b).

Proof. From (3.22), it is clear that α is a lower bound to the minimum value in (3.23). But $F^*(\cdot;0) = \widehat{F}^*(\cdot)$, as defined in (3.11) and (3.17), satisfies both constraints (3.24) and yields α as the value for the expression in (3.23). Hence, α is the minimum value and not only a lower bound. Furthermore, standard manipulations show that $\widehat{F}$ satisfies (3.24b) for all $t \in [0, T]$ if (3.25) holds. □

Remark 3.19. *Direct calculations reveal that Condition (3.25) holds for the exponential controllability assumption as introduced in Assumption 3.6 in Section 3.2.3.*

Other suboptimality estimates similar to Theorem 3.16 can be obtained using different techniques and without solving an infinite-dimensional program. For instance, simpler techniques are used in Reble and Allgöwer (2011) for calculating γ such that (3.15) holds. These techniques are similar to the ones employed by Grimm et al. (2005) for discrete-time systems. In the following, we briefly summarize results for continuous-time systems obtained without solving the infinite-dimensional program, but instead using techniques similar to Grimm et al. (2005).

Lemma 3.20 (Calculation of $\Xi^\dagger$). *Suppose that Assumptions 3.1–3.4 and Controllability Assumption 3.5 are satisfied. Then, $J_T^*(x(\delta)) \leq \Xi^\dagger \int_\delta^T F^*(t';0)dt'$, in which $\Xi^\dagger = 1 + \frac{B(T)}{T-\delta}$.*

Proof. Using (3.8a) in Lemma 3.12, we obtain

$$J_T^*(x(\delta)) \leq \min_{t^* \in [\delta,T]} \int_\delta^{t^*} F^*(t';0)dt' + B(T+\delta-t^*)F^*(t^*;0)$$

$$\leq \int_\delta^T F^*(t';0)dt' + B(T) \min_{t^* \in [\delta,T]} F^*(t^*;0)$$

$$\leq \int_\delta^T F^*(t';0)dt' + B(T) \frac{1}{T-\delta} \int_\delta^T F^*(t';0)dt'.$$

Here, we used that $F^*(t';0) \geq 0$ and the non-decreasing property of B. □

Lemma 3.21 (Calculation of $\gamma^\dagger$). *Suppose that Assumptions 3.1–3.4 and Controllability Assumption 3.5 are satisfied. Then, $\int_\delta^T F^*(t';0)dt' \leq \gamma^\dagger \int_0^\delta F^*(t';0)\,dt'$, in which $\gamma^\dagger = \frac{B(T)}{\delta}$.*

Proof. Because of (3.8b) in Lemma 3.12, we obtain

$$\int_\delta^T F^*(t';0)dt' \leq \min_{t^* \in [0,\delta]} \int_{t^*}^T F^*(t';0)dt' \leq \min_{t^* \in [0,\delta]} B(T-t^*)\, F^*(t^*;0)$$

$$\leq B(T) \min_{t^* \in [0,\delta]} F^*(t^*;0) \leq \frac{B(T)}{\delta} \int_0^\delta F^*(t';0)\,dt'.$$

Here, we used again that $F^*(t';0) \geq 0$ and the non-decreasing property of B. □

It is straightforward to see that the results on stability and suboptimality of Theorem 3.16 still hold when replacing Ξ and γ with $\Xi^\dagger$ and $\gamma^\dagger$, respectively, in (3.20).

Due to the non-decreasing property of B, it is simple to show that $\Xi < \Xi^\dagger$ and $\gamma < \gamma^\dagger$. Thus, the suboptimality estimate obtained in Theorem 3.16 is indeed qualitatively better than an estimate based on $\Xi^\dagger$ and $\gamma^\dagger$, i.e.,

$$\alpha^\dagger = 1 - \gamma^\dagger (\Xi^\dagger - 1) \leq 1 - \gamma(\Xi - 1) = \alpha.$$

However, the simple techniques used to obtain "worse" estimates can be useful in some cases for which solving the corresponding infinite-dimensional program is not possible. One example is the extension towards time-delay systems in Section 4.5, for which a modified controllability assumption, which takes the time-delay into account, is required. In this case the techniques from Lemmata 3.14 and 3.15 cannot be used to obtain suitable estimates, but the techniques from Lemmata 3.20 and 3.21 are applicable.

A similar result concerning the suboptimality estimates for discrete-time systems is proven in Worthmann (2012a,b). It is shown that the suboptimality estimate obtained in Grüne (2009); Grüne et al. (2010a) is tighter than its counterpart in Tuna et al. (2006), which has improved the results of Grimm et al. (2005). This result together with the connection reported in (Worthmann, Reble, Grüne, and Allgöwer, 2012) between our continuous-time results and the discrete-time results underpins our previous observation. Furthermore, it suggests that a continuous-time version of Tuna et al. (2006) would also yield less accurate suboptimality estimates than Theorem 3.16.

3.2.3 Special Case: Exponential Controllability

In this section, we give analytical expressions of the suboptimality estimate α for the special case of the following exponential controllability assumption, which was also considered for discrete-time systems in Grüne and Pannek (2011); Grüne et al. (2010a).

Assumption 3.6 (Exponential Controllability). *For all $x_0 \in \mathbb{R}^n$, there exists a piece-wise continuous input trajectory $\widehat{u}(\cdot\,; x_0, 0)$ with $\widehat{u}(t; x_0, 0) \in \mathbb{U}$ for all $t \in \mathbb{R}_{\geq 0}$ and corresponding state trajectory $\bar{x}_{\widehat{u}}$ such that*

$$F(\bar{x}_{\widehat{u}}(t; x_0, 0), \widehat{u}(t; x_0, 0)) \leq C\, e^{-\lambda t}\, F(x_0, 0)\,, \ \forall t \in \mathbb{R}_{\geq 0}$$

with overshoot constant $C \geq 1$ and decay rate $\lambda > 0$.

Assumption 3.6 directly implies that Assumption 3.5 holds with $B(T) = \frac{C}{\lambda}\left(1 - e^{-\lambda T}\right)$. This observation enables us to derive an analytical expression for the suboptimality estimate α.

Calculation of the Suboptimality Estimate α

Using $B(T) = \frac{C}{\lambda}\left(1 - e^{-\lambda T}\right)$ for the exponential controllability assumption, direct calculations reveal that $\int^T \frac{1}{B(t^\star)}\, dt^\star = \frac{1}{C} \ln\left(e^{\lambda T} - 1\right) + \mathfrak{C}$ with $\mathfrak{C} \in \mathbb{R}$. Hence, we can calculate Ξ and γ, see Lemmata 3.14 and 3.15, as follows

$$\frac{1}{\Xi} = 1 - \left(\frac{e^{\lambda \delta} - 1}{e^{\lambda T} - 1}\right)^{\frac{1}{C}} \quad \text{and} \quad \frac{1}{\gamma} = \left(\frac{1 - e^{-\lambda T}}{e^{-\lambda \delta} - e^{-\lambda T}}\right)^{\frac{1}{C}} - 1.$$

It is interesting to note the influence of the different parameters C, λ, δ, and T on the suboptimality estimate $\alpha = 1 - \gamma\,(\Xi - 1)$. First, it is directly clear that $\gamma > 0$ and $\Xi > 1$ and, consequently, $\alpha < 1$ for any finite prediction horizon. Second, if $T \to \infty$, then $\Xi \to 1$ and $\alpha \to 1$, which means that the performance of the MPC controller is arbitrarily close to the infinite horizon optimal performance if the prediction horizon T is chosen large enough. This again confirms the result of Proposition 3.17. Third, if $\delta \to 0$, then $\alpha \to -\infty$, which means that asymptotic stability of the closed-loop cannot be guaranteed for arbitrarily small sampling times. This result is counterintuitive, however, it is in good agreement with the results for sampled-data continuous-time systems of Grüne et al. (2010b,c). In order to obtain better stability estimates for small sampling times, we use a growth condition in the following paragraph in order to obtain better estimates, in particular for small sampling times.

Implications of the Growth Condition

In this section, we use a growth condition analogue to the discrete-time results in Grüne et al. (2010b,c) in order to obtain better suboptimality estimates and, hence, stability guarantees for shorter prediction horizons.

Assumption 3.7 (Growth Condition). *For all $x_0 \in \mathbb{R}^n$, there exists a piece-wise continuous input trajectory $\widehat{u}(\cdot; x_0, 0)$ with $\widehat{u}(t; x_0, 0) \in \mathbb{U}$ for all $t \in [0, T']$ and corresponding state trajectory $\bar{x}_{\widehat{u}}$ such that*

$$F(\bar{x}_{\widehat{u}}(t; x_0, 0), \widehat{u}(t; x_0, 0)) \le e^{\lambda_g t} F(x_0, 0) \qquad \forall t \in \mathbb{R}_{\ge 0},$$

in which $\lambda_g \in \mathbb{R}$ is a constant growth rate.

Comparing the growth condition with the exponential controllability assumption, several differences can be observed. First, λ_g can be positive. In this case, the optimal cost need not be bounded and asymptotic stability of the closed-loop with MPC cannot be guaranteed by taking only the growth condition into account. On the other hand, the growth condition gives better information for small times because no overshoot is considered. Hence, using the growth condition, we can replace $B(T)$ used in the previous paragraph by

$$B(T) = \min \left\{ \frac{C}{\lambda} \left(1 - e^{-\lambda T}\right), \frac{1}{\lambda_g} \left(e^{\lambda_g T} - 1\right) \right\}.$$

Unfortunately, it is not possible to give (simple) analytic solutions as in the previous paragraph. However, numerical solutions can be easily obtained as only integration of scalar variables is necessary for the calculation of γ and Ξ.

Comparison of different Suboptimality Estimates and Connection to Discrete-Time Results

In Figure 3.3, the suboptimality estimate α is depicted in dependence on the parameters C, λ, T, and δ. In each subfigure, the value of only one parameter is varied and the other parameters are kept constant at their default value. The default values are chosen as $C = \lambda = 2$, $\lambda_g = 0.2$, $\delta = 0.01$, and $T = 3$. As expected, α is strictly monotonically increasing in both T and λ and strictly monotonically decreasing in C. Furthermore, all estimates are improved by using the growth condition. The most significant change is for the case of $\delta \to 0$. Whereas the estimates for α tend to $-\infty$ without growth condition, the estimates with growth condition show that the closed-loop is indeed asymptotically stable for arbitrarily small sampling times.

The estimates for continuous-time systems derived in this work are in all cases better than the result of (Grüne et al., 2010c, Theorem 2) based on a sampled-data approach.

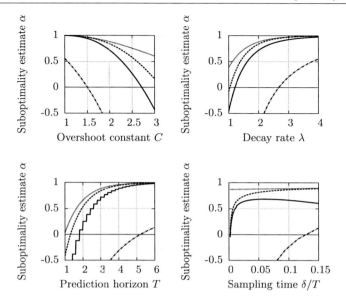

Figure 3.3: Suboptimality estimate α for the exponential controllability assumption. Solid lines: estimates from (Grüne et al., 2010c, Theorem 2); dashed lines: estimates without growth condition; dotted lines: improved estimates using the growth condition, dash-dotted lines: estimate $\alpha^\dagger$ using simpler techniques.

This indicates a gap between our continuous-time results and the discrete-time results despite the similarity of the assumptions used in the different results. A more detailed comparison of the results obtained in this section with the sampled-data implementation is given in Worthmann, Reble, Grüne, and Allgöwer (2012). The key ingredient used in this reference is to decouple the sampling time and the discretization time. This approach requires the concept of multistep feedback laws as introduced, e.g., in Grüne (2009). It is proven that the continuous-time performance estimate is always an upper bound on the performance estimate for the discrete-time results applied in a sampled-data context, but can be obtained as a limiting case for the discretization time tending to zero. For a more thorough discussion of these aspects, we refer to Worthmann, Reble, Grüne, and Allgöwer (2012).

Additionally, it can be seen that the suboptimality estimate $\alpha^\dagger$ based on Lemmata 3.20 and 3.21 is significantly worse than the other estimates. This underpins the statement of Theorem 3.18 and Remark 3.19.

3.2.4 Numerical Example

Consider the nonlinear system

$$\dot{x}_1(t) = u_1(t), \qquad \dot{x}_2(t) = u_2(t), \qquad \dot{x}_3(t) = x_1(t)\,u_2(t) - x_2(t)\,u_1(t), \qquad (3.26)$$

which is called nonholonomic integrator or Brockett integrator (Brockett, 1983). The system is not asymptotically stabilizable by continuous time-invariant state feedback and, in particular, the Jacobi linearization of the system is not asymptotically stabilizable. Therefore, the design of a corresponding control Lyapunov function is a difficult task and, indeed, there does not exist a continuously differentiable control Lyapunov function. In contrast, it is rather simple to construct an open-loop control which steers the system from any initial condition to the origin in finite time. Straightforward calculations show that the open-loop control input $\widehat{u}$ defined as

$$\widehat{u}(t) = \begin{pmatrix} -x_1(0)/t_1 \\ -x_2(0)/t_1 \end{pmatrix}, \qquad \text{for } 0 \le t < t_1$$

$$\widehat{u}(t) = \begin{pmatrix} \text{sign}(x_3(0)) \frac{\sqrt{2\pi|x_3(0)|}}{t_2} \sin(2\pi t/t_2) \\ \frac{\sqrt{2\pi|x_3(0)|}}{t_2} \cos(2\pi t/t_2) \end{pmatrix}, \qquad \text{for } t_1 \le t \le t_1 + t_2$$

steers the system to $x_1(t_1) = x_2(t_1) = 0$, $x_3(t_1) = x_3(0)$ and to $x(t_1 + t_2) = 0$. When choosing the stage cost $F(x, u) = x_1^2 + x_2^2 + \nu_3|x_3| + u_1^2 + u_2^2$, $\nu_3 \in \mathbb{R}_{>0}$, it is also possible to show that the smallest cost for such a control input $\widehat{u}$ is achieved for $t_1^* = \sqrt{3}$ and $t_2^* = \sqrt{\frac{12\pi^2 + 3\pi\nu_3}{2\pi + 3\nu_3}}$ and

$$J_{T'}^*(x_0) \le J_{T'}(x_0, \widehat{u}) \le \left(t_1^* + \frac{3 + 2\pi\nu_3}{6\pi\nu_3} t_2^* + \frac{4\pi + \nu_3}{2\nu_3 t_2^*} + \frac{1}{\pi} \right) F(x_0, 0)$$

for all $T' \in \mathbb{R}_{>0}$. This implies that stability can be guaranteed for shorter prediction horizons if a larger ν_3 is chosen. Assumption 3.5 is satisfied with $B(T') \equiv 5.71$ for $\nu_3 = 1$ and with $B(T') \equiv 4.09$ for $\nu_3 = 3$, respectively. With this information, we can guarantee asymptotic stability of the closed-loop with sampling time $\delta = 0.1$ for $T \ge 23.3$ and $T \ge 15.4$, respectively. However, it is clear that this is an overly conservative choice for $B(T')$, particularly for small T'. A better result can be obtained by taking into account that every state is an equilibrium for $u_s = 0$. Hence, we can use $B(T') = \min\{T', 5.71\}$ and $B(T') = \min\{T', 4.09\}$ to guarantee stability for $T \ge 5.71$ and $T \ge 4.09$, respectively. Furthermore, less conservative estimates on a stabilizing prediction horizon can be made by taking additional information on the controllability of the system into account.

Figure 3.4 shows simulation results for initial condition $x(0) = [3, 1, 1]^T$ and different choices of the prediction horizon T and the weight ν_3. The simulation results confirm that the closed-loop is not asymptotically stable for $T = 0.7$ and $\nu_3 = 3$ because the state x_3 does not converge to zero. On the contrary, simulation studies suggest that the closed-loop is asymptotically stable for $T = 0.8$, which reveals the conservatism of the estimates of the minimal stabilizing prediction horizon of $T \approx 4.09$. But it is interesting to note that the simulation results support the theoretical result that a larger ν_3, i.e., a larger weight on the state x_3, is beneficial for stability. This is exemplarily demonstrated in Figure 3.4 by the case of $T = 0.8$ and $\nu_3 = 1$, for which the closed-loop is not asymptotically stable, in

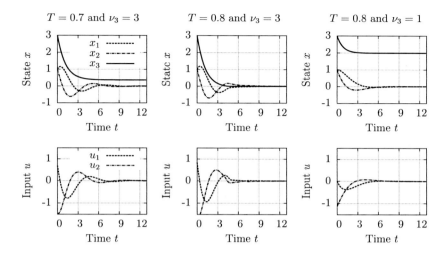

Figure 3.4: Simulation results for system (3.26) in Section 3.2.4 and unconstrained MPC for different prediction horizons T and weights ν_3.

contrast to the case of $T = 0.8$ and $\nu_3 = 3$. This shows that despite giving conservative estimates on the minimal stabilizing prediction horizon, the conditions in this chapter might give design guidelines for a suitable stage cost such stability is achieved with short prediction horizons. Similar remarks have been made for the discrete-time formulation of unconstrained MPC in (Grüne, 2009, Section 7).

3.2.5 Summary

In this section, we extended the results on model predictive control without terminal constraints and without terminal cost functions from the discrete-time case to continuous-time systems. Performance estimates and stability conditions in terms of the prediction horizon and the sampling time have been derived using an asymptotic controllability assumption and the special case of an exponential controllability assumption. These estimates, based on the solution of an infinite-dimensional program, have then been compared to other estimates. Connections of the results in this section to the previous results in discrete-time (Grüne, 2009; Grüne et al., 2010a) are thoroughly discussed in (Worthmann, Reble, Grüne, and Allgöwer, 2012).

In the following sections, we derive stability conditions for alternative MPC schemes with the goal to relax assumptions and/or guarantee stability with a shorter prediction horizon.

3.3 Unconstrained MPC with General Terminal Cost Functions

In this section, we consider a more general MPC setup without terminal constraints, allowing for a general terminal cost function. More precisely, we derive sufficient stability conditions for unconstrained MPC with a positive semi-definite terminal cost function. This result is related to the discrete-time results in Tuna et al. (2006), which takes properties of the terminal cost function into account in order to improve the stability conditions obtained in Grimm et al. (2005). However, the results reported in this section exhibit two distinguishing features in comparison to these previous results: Better estimates for the stabilizing prediction horizon are obtained and our analysis allows to recover the results for unconstrained MPC without terminal cost derived in the preceding section. In contrast, setting the terminal cost equal to zero in the result of Tuna et al. (2006) does not allow to guarantee asymptotic stability even for an arbitrarily large prediction horizon. More precisely, our result contains two previous results as special cases: First, if the terminal cost function is chosen as zero, we recover the conditions of Section 3.2 on the length of the prediction horizon such that asymptotic stability is guaranteed. Second, if the terminal cost is a control Lyapunov function conform to the stage cost, stability follows independent of the length of the prediction horizon. If the terminal cost is not a control Lyapunov function, but satisfies a significantly relaxed condition, then our results yield improved estimates for the necessary prediction horizon. Hence, the analysis in this section allows to bridge the gap between two MPC approaches, namely MPC using global control Lyapunov functions as terminal cost (Jadbabaie et al., 2001b) and unconstrained MPC without terminal cost (Grüne, 2009), which have been mostly considered separately in the literature so far, see also Figure 3.1.

3.3.1 Unconstrained MPC Setup with General Terminal Cost

We consider an MPC setup similar to the one presented in Section 3.2 with an additional positive semi-definite terminal cost function E. Therefore, the open-loop finite horizon optimal control problem to be solved at sampling instant t_i given measured state $x(t_i)$ is formulated as follows.

Problem 3.22.

$$\underset{\bar{u}\in\mathcal{PC}([t_i,t_i+T],\mathbb{R}^m)}{\text{minimize}} \quad J_{T,E}(x(t_i),\bar{u}) \tag{3.27a}$$

subject to

$$\dot{\bar{x}}(t';x(t_i),t_i) = f(\bar{x}(t';x(t_i),t_i),\bar{u}(t')), \qquad t'\in[t_i,t_i+T], \tag{3.27b}$$

$$\bar{x}(t_i;x(t_i),t_i) = x(t_i), \tag{3.27c}$$

$$\bar{u}(t')\in\mathbb{U}, \qquad t'\in[t_i,t_i+T], \tag{3.27d}$$

in which

$$J_{T,E}(x(t_i),\bar{u}) = \int_{t_i}^{t_i+T} F(\bar{x}(t';x(t_i),t_i),\bar{u}(t'))\,dt' + E(\bar{x}(t_i+T;x(t_i),t_i),\bar{u}(t')). \tag{3.27e}$$

In analogy to the notation in the previous section, we denote the optimal control input and the associated predicted state trajectory by $u_T^*(t'; x(t_i), t_i)$ and $x^*(t'; x(t_i), t_i)$, $t' \in [t_i, t_i + T]$, respectively. The optimal cost is denoted by $J_{T,E}^*(x(t_i))$. For a given sampling time δ with $0 < \delta \leq T$, the control input to the system is defined in the usual continuous-time receding horizon fashion.

Algorithm 3.23 (Unconstrained Model Predictive Control with General Terminal Cost). *At each sampling instant $t_i = i\delta$, $i \in \mathbb{N}_0$, measure the state $x(t_i)$ and solve Problem 3.22. Apply the input*

$$u_{\mathrm{MPC}}(t) = u_T^*(t; x(t_i), t_i), \quad t_i \leq t < t_i + \delta \tag{3.28}$$

to the system until the next sampling instant $t_{i+1} = t_i + \delta$.

Remark 3.24. *A slightly different approach for unconstrained MPC with terminal weights is considered in (Grüne and Pannek, 2011, Section 7.2), in which a scalar terminal weighting on the stage cost is introduced for discrete-time MPC.*

3.3.2 Asymptotic Stability

We use the following assumption regarding the stage cost F and the terminal cost E.

Assumption 3.8. *The stage cost $F : \mathbb{R}^n \times \mathbb{U} \to \mathbb{R}_{\geq 0}$ and the terminal cost function $E : \mathbb{R}^n \to \mathbb{R}_{\geq 0}$ are continuous, $F(0,0) = E(0) = 0$, and there is a class $\mathcal{K}_\infty$ function $\underline{\alpha}_F : \mathbb{R}_{\geq 0} \to \mathbb{R}_{\geq 0}$ such that for all $x \in \mathbb{R}^n$ and for all $u \in \mathbb{U}$*

$$F(x, u) \geq F(x, 0) \geq \underline{\alpha}_F(|x|) \tag{3.29a}$$

$$and \quad E(x) \geq 0. \tag{3.29b}$$

Note that the terminal cost E is not necessarily positive definite, but only positive semi-definite, which allows to consider the unconstrained MPC setup presented in Section 3.2 as a special case of the current setup for $E \equiv 0$. This is in contrast to the classical MPC framework with guaranteed stability, which was recalled in Section 2.1.3. In particular, the extended Lyapunov inequality in Assumption 2.5 can only be satisfied for all states $x \in \mathbb{R}^n$ (which is necessary because we do not consider terminal state constraints) if the terminal cost E is positive definite or with some minor modifications for the trivial case of an open-loop stable system with a stage cost which does not depend on the state, i.e., $F(x, u) = F(0, u)$ for all $x \in \mathbb{R}^n$. The extended Lyapunov inequality in Assumption 2.5 is in the following replaced by a significantly less restrictive controllability assumption, see Assumption 3.10. This assumption can also be satisfied in the case of $E \equiv 0$, but possibly improves the estimates on a minimal stabilizing prediction horizon.

In the following analysis, similar to Section 3.2.2, we consider the two consecutive sampling instants $t_0 = 0$ and $t_1 = \delta$ without loss of generality. With slight abuse of notation, we use the following abbreviations

$$F^*(t; t_i) = F(x^*(t; x(t_i), t_i), u_T^*(t; x(t_i), t_i)), \tag{3.30a}$$

$$E^*(t; t_i) = E(x^*(t; x(t_i), t_i)), \tag{3.30b}$$

for $t \in [t_i, t_i + T]$ and $t_i \in \{0, \delta\}$. From this definition, it directly follows that

$$J_{T,E}^*(x(t_i)) = \int\limits_{t_i}^{t_i+T} F^*(t'; t_i)dt' + E^*(t_i + T; t_i) \,. \tag{3.31}$$

In order to guarantee stability, we use the following two assumptions.

Assumption 3.9 (Controllability Assumption based on F). *For all $T' \in \mathbb{R}_{\geq 0}$ and $x_0 \in \mathbb{R}^n$, there exists a piece-wise continuous input trajectory $\widehat{u}(\cdot; x_0, 0)$ with $\widehat{u}(t; x_0, 0) \in \mathbb{U}$ for all $t \in [0, T']$ and*

$$J_{T',E}^*(x_0) \leq J_{T',E}(x_0, \widehat{u}) \leq B_E(T')F(x_0, 0) \,,$$

in which $B_E : \mathbb{R}_{\geq 0} \to \mathbb{R}_{>0}$ is a continuous and bounded function.

Assumption 3.10 (Controllability Assumption based on E). *For all $x_0 \in \mathbb{R}^n$, there exists a piece-wise continuous input trajectory $\widehat{u}(\cdot; x_0, 0)$ with $\widehat{u}(t; x_0, 0) \in \mathbb{U}$ for all $t \in [0, \delta]$ and such that*

$$\Gamma_E \left(\int\limits_0^\delta F(\bar{x}(t'; x_0, 0), \widehat{u}(t'; x_0, 0))dt' + E(\bar{x}(\delta; x_0, 0)) \right) \leq E(x_0) \,, \tag{3.32}$$

in which $\Gamma_E \in [0, 1]$.

Assumption 3.9 is a standard assumption in unconstrained MPC, see also Assumption 3.5 in Section 3.2 and (Grimm et al., 2005; Grüne, 2009; Grüne et al., 2010a). However, note that B_E in Assumption 3.9 depends explicitly on the terminal cost E due to the definition of $J_{T,E}$. Hence, it might be only satisfied for larger values of B_E compared to B used in Assumption 3.5 in Section 3.2.

Assumption 3.10 is always satisfied for $\Gamma_E = 0$. On the other hand, if no terminal cost term is considered, i.e., $E(x) = 0$ as in (Grüne, 2009; Grüne and Pannek, 2011; Grüne et al., 2010a) and Section 3.2, Assumption 3.10 can only be satisfied for $\Gamma_E = 0$. For $0 < \Gamma_E < 1$, Assumption 3.10 can be interpreted as the terminal cost $E(x)$ being "similar" to a CLF, however, in a significantly weaker sense as shown in the following proposition.

Proposition 3.25 (CLF-like Condition for Assumption 3.10). *Suppose that there exists $\nu \in \mathbb{R}_{\geq 0}$ and a control law $k : \mathbb{R} \to \mathbb{U}$ such that the derivative of E along trajectories of the closed-loop consisting of system (3.1) with control input $u = k(x)$ satisfies for all $x \in \mathbb{R}^n$*

$$\dot{E}(x) \leq -F(x, k(x)) + \nu \, E(x) \,. \tag{3.33}$$

Then, Assumption 3.10 is satisfied for $\Gamma_E = e^{-\nu \delta}$.

Proof. Integrating Inequality (3.33) from $t = 0$ to $t = \delta$ yields

$$\int\limits_0^\delta F(\bar{x}(t'; x_0, 0), k(\bar{x}(t'; x_0, 0)))dt' + E(\bar{x}(\delta; x_0, 0)) \leq E(x_0) + \nu \int\limits_0^\delta E(\bar{x}(t'; x_0, 0)) \, dt' \,.$$

By taking $F(x, u) \geq 0$ into account, we can immediately see that Inequality (3.33) implies that $E(\bar{x}(t'; x_0, 0)) \leq e^{\nu t'} E(x_0)$. Combining these two findings allows to deduce

$$\int_0^\delta F(\bar{x}(t'; x_0, 0), k(\bar{x}(t'; x_c, 0)))dt' + E(\bar{x}(\delta; x_0, 0)) \leq E(x_0) + \nu \int_0^\delta e^{\nu t'} E(x_0) \, dt'$$

$$= e^{\nu \delta} E(x_0).$$

Hence, Inequality (3.32) is satisfied for $\Gamma_E = e^{-\nu \delta}$, which completes the proof. □

For the special case $\nu = 0$, Condition (3.33) in Proposition 3.25 corresponds to E being an F-conform control Lyapunov function (CLF), which is a common assumption in MPC in order to guarantee stability of the closed-loop, see, e.g., the well-established general framework recalled in Section 2.1.3 and (Chen and Allgöwer, 1998; Fontes, 2001; Jadbabaie et al., 2001b; Mayne et al., 2000). In this case, we obtain $\Gamma_E = 1$ and Assumption 3.10 is an integrated, slightly more general, version of the the extended Lyapunov inequality $\dot{E}(x) \leq -F(x, k(x))$, see Assumption 2.5 in Section 2.1.3. Note the similarity to Condition (4.29) in Assumption 4.8 used for the Razumikhin-based design of a terminal cost for nonlinear time-delay systems in Section 4.3.3.

For the general case $\nu > 0$, this Lyapunov condition is relaxed and, consequently, additional arguments are required in order to guarantee stability. In this section, a prediction horizon chosen "sufficiently large" is considered with explicit conditions on the prediction horizon. Note that these conditions are closely related to the results of Section 3.2 and are based on Assumption 3.9. To this end, we can show three intermediate results proven in the following using the two assumptions above. The main result of this section is then summarized in Theorem 3.29.

Lemma 3.26 (Calculation of Ξ_E). *Suppose that Assumptions 3.1–3.3, Assumption 3.8, and Controllability Assumption 3.9 are satisfied. Then,*

$$J_{T,E}^*(x(\delta)) \leq \Xi_E \int_\delta^T F^*(t'; 0)dt', \tag{3.34}$$

in which

$$\frac{1}{\Xi_E} = 1 - \exp\left(-\int_\delta^T \frac{1}{B_E(T + \delta - t^*)} \, dt^*\right). \tag{3.35}$$

Proof. The proof is similar to the proof of Lemma 3.14 for unconstrained MPC without terminal cost. Controllability Assumption 3.9 implies that

$$J_{T,E}^*(x(\delta)) \leq \int_\delta^{t^*} F^*(t'; 0)dt' + B_E(T + \delta - t^*)F^*(t^*; 0) \tag{3.36}$$

holds for all $t^* \in [\delta, T]$. Consider any piece-wise continuous function $F^*(t; 0) : [\delta, T] \to \mathbb{R}_{\geq 0}$ satisfying (3.36) and define the function $\widehat{F}^*(t) : [\delta, T] \to \mathbb{R}_{\geq 0}$ by

$$\widehat{F}^*(t) = \frac{J_{T,E}^*(x(\delta))}{B_E(T + \delta - t)} e^{-\int_\delta^t \frac{1}{B_E(T + \delta - t')} dt'}. \tag{3.37}$$

In the next step, we show that for all piece-wise continuous functions $F^*(t; 0)$ satisfying (3.36), the following holds

$$\int_{\delta}^{T} \widehat{F}^*(t')dt' \leq \int_{\delta}^{T} F^*(t'; 0)dt' . \tag{3.38}$$

To this end, note that $\widehat{F}^*(t)$ satisfies (3.36) with equality (instead of inequality) for all $t^* \in [\delta, T]$, which can be shown by direct evaluation. For the sake of contradiction, assume $\int_{\delta}^{T} \widehat{F}^*(t')dt' > \int_{\delta}^{T} F^*(t'; 0)dt'$. But then there exists a $t \in [\delta, T]$ for which

$$\int_{\delta}^{t} \widehat{F}^*(t')dt' \geq \int_{\delta}^{t} F^*(t'; 0)dt' \quad \text{and} \quad \widehat{F}^*(t) > F^*(t; 0) .$$

But this contradicts (3.36), which shows that (3.38) holds. On the other hand, direct calculations reveal

$$J_{T,E}^*(x(\delta)) = \Xi_E \int_{\delta}^{T} \widehat{F}^*(t')dt' \tag{3.39}$$

for Ξ_E defined in (3.35). Combining (3.38) and (3.39) yields (3.34). $\qquad \square$

Lemma 3.27 (Calculation of γ_E). *Suppose that Assumptions 3.1–3.3, Assumption 3.8, and Controllability Assumption 3.9 are satisfied. Then,*

$$\int_{\delta}^{T} F^*(t'; 0)dt' + E^*(T; 0) \leq \gamma_E \int_{0}^{\delta} F^*(t'; 0)\, dt' , \tag{3.40}$$

in which

$$\frac{1}{\gamma_E} = \exp\left(\int_{0}^{\delta} \frac{1}{B_E(T - t^*)}\, dt^* \right) - 1 . \tag{3.41}$$

Proof. The proof mirrors the proof of Lemma 3.15 for unconstrained MPC without terminal cost. Controllability Assumption 3.9 and the principle of optimality imply that

$$\int_{t^*}^{T} F^*(t'; 0)dt' + E^*(T; 0) \leq B_E(T - t^*)\, F^*(t^*; 0) \tag{3.42}$$

holds for all $t^* \in [0, T]$. Consider any piece-wise continuous function $F^*(t; 0) : [0, T] \to \mathbb{R}_{\geq 0}$ satisfying (3.42) and define the function $\widehat{F}^*(t) : [0, \delta] \to \mathbb{R}_{\geq 0}$ by

$$\widehat{F}^*(t) = \frac{\int_{\delta}^{T} F^*(t'; 0)dt' + E^*(T; 0)}{B_E(T - t)}\, e^{-\int_{\delta}^{t} \frac{1}{B_E(T - t')}\, dt'} . \tag{3.43}$$

In the next step, we show that for all piece-wise continuous functions $F^*(t;0)$ satisfying (3.42), the following holds

$$\int_0^\delta \widehat{F}^*(t')dt' \leq \int_0^\delta F^*(t';0)dt'. \tag{3.44}$$

To this end, note that the anti-derivative of $\widehat{F}^*$ is

$$\int^t \widehat{F}^*(t')dt' = -\left(\int_\delta^T F^*(t';0)dt' + E^*(T;0)\right) e^{-\int_\delta^t \frac{1}{B_E(T-t')}\,dt'} + \mathfrak{C}$$

with $\mathfrak{C} \in \mathbb{R}$. Hence, $\widehat{F}^*$ satisfies (3.42) with equality for all $t^* \in [0,\delta]$ in the sense of

$$\int_{t^*}^\delta \widehat{F}^*(t')dt' + \int_\delta^T F^*(t';0)dt' + E^*(T;0) = B_E(T-t^*)\,\widehat{F}^*(t^*).$$

For any $F^*(t;0)$ satisfying (3.42), we know that $F^*(\delta;0) - \widehat{F}^*(\delta) \geq 0$ and

$$\int_{t^*}^\delta \left(F^*(t';0) - \widehat{F}^*(t')\right) dt' \leq B_E(T-t^*)\left(F^*(t^*;0) - \widehat{F}^*(t^*)\right)$$

holds for all $t^* \in [0,\delta]$. Define $\mathfrak{F}(t^*) = F^*(\delta - t^*;0) - \widehat{F}^*(\delta - t^*)$, for which $\mathfrak{F}(0) \geq 0$ and $\int_0^{t^*} \mathfrak{F}(t')dt' \leq B_E(T - \delta + t^*)\mathfrak{F}(t^*)$. Due to the comparison lemma (Khalil, 2002), it follows that $\mathfrak{F}(t^*) \geq 0$ for all $t^* \in [0,\delta]$. Consequently, Equation (3.44) holds. On the other hand, direct calculations show that

$$\int_0^\delta \widehat{F}^*(t')\,dt' = \frac{1}{\gamma_E}\left(\int_\delta^T F^*(t';0)dt' + E^*(T;0)\right), \tag{3.45}$$

with γ_E defined by (3.41). Finally, the combination of (3.44) and (3.45) implies (3.40). This completes the proof. $\qquad\square$

Lemma 3.28 (Direct Consequence of Controllability Assumption 3.10). *Suppose that Assumptions 3.1–3.3, Assumption 3.8, and Controllability Assumption 3.10 are satisfied. Then,*

$$\Gamma_E\,J_{T,E}^*(x(\delta)) \leq \Gamma_E \int_\delta^T F^*(t';0)dt' + E^*(T;0). \tag{3.46}$$

Proof. The proof follows directly from optimality of $J_{T,E}^*(x(\delta))$ and using Assumption 3.10 for $x_0 = x^*(T;x(0),0)$, which provides an upper bound on the cost on the interval $[T, T + \delta]$. $\qquad\square$

We are now able to state our main result on stability of unconstrained MPC with general positive semi-definite terminal cost functions.

Theorem 3.29 (Stability of Unconstrained MPC with General Terminal Cost Function). *Suppose that Assumptions 3.1–3.3, Assumption 3.8, and Controllability Assumptions 3.9 and 3.10 are satisfied for system* (3.1) *and consider*

$$\alpha_E = 1 - \gamma_E \left(\frac{\Xi_E}{1 + (\Xi_E - 1)\Gamma_E} - 1 \right), \tag{3.47}$$

with γ_E and Ξ_E defined in Lemmata 3.26 and 3.27, respectively. If $\alpha_E > 0$, the closed-loop system resulting from the application of the model predictive controller according to Algorithm 3.23 to system (3.1) *is asymptotically stable.*

Proof. Multiplying Inequality (3.34) in Lemma 3.26 with $1 - \Gamma_E$ and multiplying Inequality (3.46) in Lemma 3.28 with Ξ_E, and adding the two resulting inequalities yields

$$J_{T,E}^*(x(\delta)) \leq \frac{\Xi_E}{1 + (\Xi_E - 1)\Gamma_E} \left(\int_{\delta}^{T} F^*(t';0)dt' + E^*(T;0) \right).$$

Following the proof of Theorem 3.16, using (3.31), and Lemma 3.27, we obtain

$$J_{T,E}^*(x(\delta)) - J_{T,E}^*(x(0))$$

$$\leq \left(\frac{\Xi_E}{1 + (\Xi_E - 1)\Gamma_E} - 1 \right) \left(\int_{\delta}^{T} F^*(t';0)dt' + E^*(T;0) \right) - \int_{0}^{\delta} F^*(t';0)\,dt'$$

$$\overset{(3.40)}{\leq} \underbrace{\left(\gamma_E \left(\frac{\Xi_E}{1 + (\Xi_E - 1)\Gamma_E} - 1 \right) - 1 \right)}_{=-\alpha_E} \int_{0}^{\delta} F^*(t';0)dt'. \tag{3.48}$$

In analogy to the proof of Theorem 3.16, asymptotic stability follows directly from standard arguments in optimal control, the lower bounds on the stage cost and the terminal cost (3.29), and application of Barbalat's Lemma (Barbalat, 1959; Khalil, 2002). $\square$

Remark 3.30. *In contrast to the unconstrained MPC schemes without terminal cost, see, e.g., Grüne (2009); Grüne et al. (2010a) and Section 3.2, $J_{T,E}^*$ is not necessarily monotonically increasing in T. Hence, α_E does not give a suboptimality estimate of the closed-loop compared to the infinite horizon optimal controller in contrast to α in Theorem 3.16.*

Remark 3.31. *As in Proposition 3.17, $\alpha_E > 0$, and thereby also asymptotic stability, can always be guaranteed for a prediction horizon chosen large enough because $\Xi_E \to 1$ and $\alpha_E \to 1$ for $T \to \infty$.*

For $\Gamma_E = 0$, for which Assumption 3.10 is trivially satisfied, we recover the stability condition from Theorem 3.16 and consequently the results for unconstrained MPC without terminal cost. These results have been shown to be the "best possible" stability conditions based only on the Controllability Assumption 3.9, see Theorem 3.18. Here, "best possible" refers to the largest α_E for which (3.48) holds. Loosely speaking, this can be seen because $\widehat{F}^*$, defined in the proofs of Lemmata 3.26 and 3.27, satisfies all conditions implied by Controllability Assumption 3.9 and yields $\alpha_E = 1 - \gamma_E(\Xi_E - 1)$ for (3.48). Hence, one can only guarantee (3.48) for a larger α_E by taking into account additional information, e.g., about the system or the optimal cost function.

If Assumption 3.9 is satisfied for $\Gamma_E = 1$, for instance if E is a global F-conform control Lyapunov function, then asymptotic stability is guaranteed independently of the Controllability Assumption 3.9. Thus, we recover the previous stability result using global control Lyapunov functions (Jadbabaie et al., 2001b).

Summarizing, Theorem 3.29 allows in some sense to bridge the gap between the stability results for MPC schemes using control Lyapunov functions as terminal cost and the more recently developed MPC schemes based on controllability assumptions. However, the case of Assumption 3.10 only being satisfied in a terminal region around the origin has not been treated so far. An additional terminal constraint makes the verification of Assumption 3.9 significantly harder, which underpins the need for a different MPC formulation in this case. Two possible MPC formulations relying only on local controllability assumptions are introduced in Sections 3.5 and 3.6.

3.3.3 Numerical Example

Consider the nonlinear system

$$\dot{x}_1(t) = u(t), \qquad \dot{x}_2(t) = u(t)^3, \qquad (3.49)$$

which is called cubic integrator (Grimm et al., 2005). Similar to the Brockett integrator (3.26) in Section 3.2.4, system (3.49) is not asymptotically stabilizable by continuous time-invariant state feedback and, in particular, the Jacobi linearization of the system is not asymptotically stabilizable. Therefore, the design of a corresponding control Lyapunov function is a difficult task and, indeed, there does not exist a continuously differentiable control Lyapunov function. In contrast, it is rather simple to construct an open-loop control which steers the system from any initial condition to the origin in finite time. Straightforward calculations show that the open-loop control input $\widehat{u}$ defined below steers the system to $x(4) = 0$, see also (Grimm et al., 2005),

$$
\begin{aligned}
\widehat{u}(t) &= -x_1(0) && \text{for } 0 \leq t < 1, & \widehat{u}(t) &= a\Psi && \text{for } 1 \leq t < 2, \\
\widehat{u}(t) &= b\Psi && \text{for } 2 \leq t < 3, & \widehat{u}(t) &= \Psi && \text{for } 3 \leq t < 4,
\end{aligned}
$$

in which $a = -\frac{1}{2} + \sqrt{\frac{7}{12}}$, $b = -\frac{1}{2} - \sqrt{\frac{7}{12}}$, and $\Psi = (x_2(0) - x_1(0)^3)^{1/3}$. Moreover, every state is an equilibrium for $u_s = 0$. When choosing the stage cost $F(x, u) = x_1^6 + x_2^2 + u^6$, the terminal cost $E(x) = \nu(x_1^6 + x_2^2)$ with $\nu \in \mathbb{R}_{\geq 0}$, and the sampling time $\delta = 0.1$, we can show that Assumption 3.9 is satisfied for

$$B_E(T') = \min\left\{15.36 + \nu(1 - T'/4)^2, T' + \nu\right\}$$

and Assumption 3.10 is satisfied for $\Gamma_E = \frac{\nu}{B_E(\delta)}$. For $\nu = 0$, i.e., unconstrained MPC without terminal cost as considered in Section 3.2, we have $\Gamma_E = 0$ and Theorems 3.16 and 3.29 guarantee stability for a prediction horizon $T \geq 15.4$. In contrast when choosing

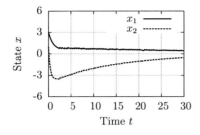 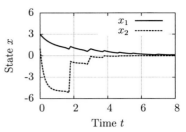

Figure 3.5: Simulation results for system (3.49) in Section 3.3.3 and unconstrained MPC with terminal cost. Left: no terminal cost $\nu = 0$; right: terminal cost with $\nu = 10$.

$\nu = 5$, we have $\Gamma_E = 0.980$ and Theorem 3.29 guarantees stability for a prediction horizon $T \geq 10.9$. The choice of $\nu = 10$ yields $\Gamma_E = 0.9901$ and guaranteed stability for $T \geq 8.0$.

In Figure 3.5, we demonstrate these findings with simulation results. We have chosen the initial condition $x_0 = [4, 1]^T$ and the very short prediction horizon $T = 0.3$. The closed-loop resulting from unconstrained MPC without terminal cost, i.e., $\nu = 0$, does not converge to the origin as shown on the left hand side in Figure 3.5. In contrast, $\nu = 10$ ensures convergence of the closed-loop as displayed on the right hand side.

3.3.4 Summary

In this section, we derived stability conditions for unconstrained MPC with a general positive semi-definite terminal cost. We have shown that a terminal cost satisfying a relaxed Lyapunov condition can lead to stability guarantees for shorter prediction horizons. Moreover, the results bridge the gap between MPC schemes using control Lyapunov functions as terminal cost and unconstrained MPC schemes without terminal cost. The results have been illustrated for the simple example of the cubic integrator.

3.4 Unconstrained MPC with Exponential Weighting

In this section, we propose an alternative additional weighting in the cost function, which possibly allows to guarantee stability for shorter prediction horizons. In contrast to an additional terminal cost, we consider an exponential weighting on the stage cost in the following.

3.4.1 Unconstrained MPC Setup with Exponential Weighting

The open-loop finite horizon optimal control problem with exponential weighting to be solved at sampling instant t_i given measured state $x(t_i)$ is formulated as follows.

Problem 3.32.

$$\underset{\bar{u}\in\mathcal{PC}([t_i,t_i+T],\mathbb{R}^m)}{\text{minimize}} \; J_{T,\mu}(x(t_i),\bar{u}) \tag{3.50a}$$

subject to

$$\dot{\bar{x}}(t';x(t_i),t_i) = f(\bar{x}(t';x(t_i),t_i),\bar{u}(t')), \qquad t'\in[t_i,t_i+T], \tag{3.50b}$$
$$\bar{x}(t_i;x(t_i),t_i) = x(t_i), \tag{3.50c}$$
$$\bar{u}(t')\in\mathbb{U}, \qquad t'\in[t_i,t_i+T], \tag{3.50d}$$

in which

$$J_{T,\mu}(x(t_i),\bar{u}) = \int_{t_i}^{t_i+T} \beta(t'-t_i)\,F(\bar{x}(t';x(t_i),t_i),\bar{u}(t'))\,dt' \tag{3.50e}$$

with $\beta(t) = e^{\mu t}$ for some constant $\mu\in\mathbb{R}_{\geq0}$ being an exponential weighting on the stage cost and no terminal cost terms are considered.

Remark 3.33. *The exponential weighting introduced in (3.50e) is inspired by a loosely related exponential weighting used in Reble et al. (2011a) for the definition of terminal cost terms for nonlinear time-delay systems, see also Section 4.3.4. A negative exponential weighting $\mu\in\mathbb{R}_{<0}$ is sometimes used as a discount factor in the framework of economic MPC, see, e.g., Huang et al. (2011); Würth et al. (2009).*

The minimizer of Problem 3.32 and the associated optimal cost are again denoted by $u_T^*(t';x(t_i),t_i)$ for all $t'\in[t_i,t_i+T]$ and $J_T^*(x(t_i))$, respectively. The control input to the system is defined in the usual continuous-time receding horizon fashion.

Algorithm 3.34 (Unconstrained Model Predictive Control with Exponential Weighting). *At each sampling instant $t_i = i\delta$, $i\in\mathbb{N}_0$, measure the state $x(t_i)$ and solve Problem 3.32. Apply the input*

$$u_{\mathrm{MPC}}(t) = u_T^*(t;x(t_i),t_i), \quad t_i\leq t<t_i+\delta \tag{3.51}$$

to the system until the next sampling instant $t_{i+1} = t_i+\delta$.

3.4.2 Asymptotic Stability

Following the analysis of Section 3.2, we consider the two consecutive sampling instants $t_0 = 0$ and $t_1 = \delta$ and we will use the abbreviation $F^*(t;t_i)$ introduced in (3.5). The optimal cost is denoted by $J_{T,\mu}^*(x(t_i))$ and we have by definition

$$J_{T,\mu}^*(x(t_i)) = \int_{t_i}^{t_i+T} \beta(t'-t_i)F^*(t';t_i)dt'. \tag{3.52}$$

In contrast to Sections 3.2 and 3.3, we restrict ourselves to the exponential controllability assumption with a minor addition as follows.

Assumption 3.11 (Exponential Controllability). *Assumption 3.6 in Section 3.2.3 is satisfied for a decay rate $\lambda > \mu \geq 0$.*

As already mentioned, exponential controllability is a common assumption in unconstrained MPC (Grüne, 2009; Grüne and Pannek, 2011; Grüne et al., 2010a) and a special case of Assumption 3.5, which is satisfied with $B(T) = \frac{C}{\lambda}(1 - e^{-\lambda T})$ whenever the exponential controllability assumption holds.

Similar to the results of Lemmata 3.14 and 3.15, the exponential controllability assumption allows to give upper bounds on the optimal cost over certain intervals as stated in the following lemma.

Lemma 3.35 (Calculation of Ξ_μ and γ_μ). *Suppose that Assumptions 3.1–3.4 and the Exponential Controllability Assumption 3.11 are satisfied. Then,*

$$J^*_{T,\mu}(x(\delta)) \leq \Xi_\mu \int_\delta^T \beta(t') F^*(t'; 0) dt', \tag{3.53a}$$

$$\int_\delta^T \beta(t') F^*(t'; 0) dt' \leq \gamma_\mu \int_0^\delta \beta(t') F^*(t'; 0) dt', \tag{3.53b}$$

in which Ξ_μ and γ_μ are defined by

$$\frac{e^{-\mu\delta}}{\Xi_\mu} = 1 - \left(\frac{e^{(\lambda-\mu)\delta} - 1}{e^{(\lambda-\mu)T} - 1} \right)^{\frac{1}{C}}, \tag{3.54a}$$

$$\frac{1}{\gamma_\mu} = \left(\frac{1 - e^{-(\lambda-\mu)T}}{e^{-(\lambda-\mu)\delta} - e^{-(\lambda-\mu)T}} \right)^{\frac{1}{C}} - 1. \tag{3.54b}$$

Proof. Due to (3.52) and the definition of $\beta(t)$, we have

$$J^*_{T,\mu}(x(\delta)) = \int_\delta^{\delta+T} \beta(t' - \delta) F^*(t'; \delta) dt' = e^{-\mu\delta} \int_\delta^{\delta+T} \beta(t') F^*(t'; \delta) dt'.$$

The Exponential Controllability Assumption 3.11 and optimality of $J^*_{T,\mu}(x(\delta))$ then imply that

$$e^{\mu\delta} J^*_{T,\mu}(x(\delta)) \leq \int_\delta^{t^*} \beta(t') F^*(t'; 0) dt' + B_\mu(T + \delta - t^*) \beta(t^*) F^*(t^*; 0) \tag{3.55}$$

holds for all $t^* \in [\delta, T]$ with $B_\mu(T) = \frac{C}{\lambda-\mu}(1 - e^{-(\lambda-\mu)T})$. Noting the similarity of (3.8a) and (3.55) allows to follow the proof of Lemma 3.14. Defining $\widehat{F}^*(t) : [\delta, T] \to \mathbb{R}_{\geq 0}$ by

$$\widehat{F}^*(t) = e^{-\mu(t-\delta)} \frac{J^*_{T,\mu}(x(\delta))}{B_\mu(T + \delta - t)} e^{-\int_\delta^t \frac{1}{B_\mu(T+\delta-t')} dt'}$$

allows to show $\int_\delta^T \beta(t')\widehat{F}^*(t')dt' \leq \int_\delta^T \beta(t')F^*(t';0)dt'$ using (3.55). Since $J_{T,\mu}^*(x(\delta)) = \Xi_\mu \int_\delta^T \beta(t')\widehat{F}^*(t')dt'$, we have proved (3.53a) and (3.54a).

The proof of (3.53b) and (3.54b) follows exactly the lines of the proof of Lemma 3.15 and replacing $F^*(t';0)$ and $\widehat{F}^*(t)$ by $\beta(t')F^*(t';0)$ and $\beta(t')\widehat{F}^*(t)$, respectively, in all expressions. □

Comparing these results to Section 3.2.3, or equivalently by substitution of $B(T) = \frac{C}{\lambda}\left(1 - e^{-\lambda T}\right)$ in the results of Lemmata 3.14 and 3.15, we note two differences. First, the decay rate λ is replaced by $\lambda - \mu$ in all expressions. Second, the expression for Ξ is multiplied by an additional term $e^{-\mu\delta}$. This additional factor is directly caused by shifting the prediction horizon when using an exponential weighting in the stage cost and beneficial for achieving $\alpha > 0$, and, hence, for stability guarantees.

Theorem 3.36 (Stability of Unconstrained MPC with Exponential Weighting). *Suppose that Assumptions 3.1–3.4 and Exponential Controllability Assumption 3.11 are satisfied for the nonlinear system (3.1). Furthermore, suppose that*

$$\alpha_\mu = 1 - \gamma_\mu\left(\Xi_\mu - 1\right) > 0\,, \tag{3.56}$$

with Ξ_μ and γ_μ defined in (3.54). Then, the closed-loop system resulting from the application of the model predictive controller according to Algorithm 3.34 to system (3.1) is asymptotically stable.

Proof. The proof follows closely the proofs of Theorems 3.16 and 3.29. By noting that $\Xi_\mu > 1$, we obtain

$$J_{T,\mu}^*(x(\delta)) - J_{T,\mu}^*(x(0)) \stackrel{(3.52)}{=} J_{T,\mu}^*(x(\delta)) - \int_0^T \beta(t')F^*(t';0)\,dt'$$

$$\stackrel{(3.53a)}{\leq} (\Xi_\mu - 1)\int_\delta^T \beta(t')F^*(t';0)\,dt' - \int_0^\delta \beta(t')F^*(t';0)\,dt'$$

$$\stackrel{(3.53b)}{\leq} \underbrace{(\gamma_\mu(\Xi_\mu - 1) - 1)}_{=-\alpha_\mu}\int_0^\delta \beta(t')F^*(t';0)dt'\,. \tag{3.57}$$

For $\alpha_\mu > 0$, asymptotic stability follows analogue to the proof of Theorem 3.16 when replacing J_T^* by $J_{T,\mu}^*$. □

The stability condition in the preceding theorem can be improved if a growth condition on the system is taken into account, see Assumption 3.7 in Section 3.2.3 and Grüne et al. (2010b). In particular, this additional information will allow to guarantee stability for shorter prediction horizons if a small sampling time is chosen.

Theorem 3.37. *Suppose that Assumptions 3.1–3.4, the Exponential Controllability Assumption 3.11 and Assumption 3.7 are satisfied for the nonlinear system (3.1). Then,*

Equation (3.53a) and the assertion of Theorem 3.36 also hold for Ξ_μ replaced by

$$\frac{e^{-\mu\delta}}{\Xi_\mu} = 1 - \exp\left(-\int_\delta^T \frac{1}{B_{\mu,\lambda_g}(T+\delta-t^*)}\,dt^*\right).$$

in which

$$B_{\mu,\lambda_g}(T) = \min\left\{B_\mu(T),\, B_{\lambda_g}(T)\right\},\quad B_{\lambda_g}(T) = \frac{1}{\lambda_g+\mu}\left(e^{(\lambda_g+\mu)T}-1\right).$$

Proof. Assumption 3.7, i.e., the growth condition, implies that (3.55) is also satisfied when $B_\mu(T)$ is replaced by $B_{\lambda_g}(T)$. Hence, $B_\mu(T)$ can be replaced by $B_{\mu,\lambda_g}(T)$ in the remaining parts of the proof of Lemma 3.35. This allows to follow the proof of Lemma 3.14 in order to complete the proof. □

3.4.3 Numerical Example

In order to briefly illustrate the effect of the newly introduced exponential weighting, we consider the following case:

$$C = 2.5,\quad \lambda = 4,\quad \text{and}\quad \delta = 0.15.$$

On the left hand side of Figure 3.6, the values for α given by Theorem 3.36 are shown as a function of the prediction horizon T and the parameter of the exponential weighting μ. On the right hand side of Figure 3.6, the smallest prediction horizon is shown for which asymptotic stability is guaranteed by virtue of Theorem 3.36, i.e.,

$$T^* = \inf_{T\in\mathbb{R}_{\geq 0}} T,\quad \text{s.t. } \alpha > 0.$$

In this example, we see the benefits of the additional exponential weighting with respect to the minimal prediction horizon with guaranteed stability. However, we also note that T^* does not necessarily decrease monotonically in μ. Unfortunately, the opposite can also be the case, i.e., the use of exponential weighting might require longer prediction horizons for satisfaction of the sufficient stability conditions established in the present work.

3.4.4 Summary

In this section, we presented an unconstrained MPC scheme using an additional exponential weighting term on the stage cost along the entire prediction horizon. Although only the first part of the optimal open-loop input trajectory calculated at each sampling instant is actually applied in MPC, the last part of the prediction horizon plays a crucial role with respect to stability. Similar to the use of a terminal cost function, stability guarantees for shorter prediction horizons are possible when using the exponential weighting proposed in this section. Another advantage is that this weighting allows to consider a local controllability assumption in combination with a generalized terminal constraint in order to guarantee stability, see Section 3.6 for more details.

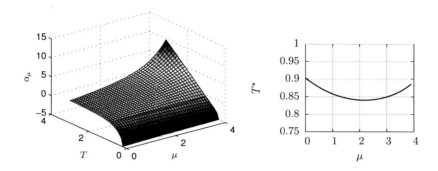

Figure 3.6: Left: α_μ from (3.56) in Theorem 3.36 as a function of the prediction horizon T and the exponential weighting parameter μ. The red line depicts the stability limit $\alpha_\mu = 0$. Right: Shortest prediction horizon T^* which guarantees $\alpha_\mu > 0$ and, therefore, closed-loop stability.

3.5 A Unifying Framework using Integral Terminal Cost Terms

As discussed in the introduction of this chapter, one can distinguish two classes of MPC schemes with guaranteed stability, CLF-MPC and unconstrained MPC. In the literature, both classes have mostly been considered separately. The results in this section provide a first step towards a unifying view on both classes of schemes. In contrast to Section 3.3, in which we have investigated the use of assumptions from both MPC schemes, we propose the use of one generalized assumption in this section. In our general framework, we identify three main ingredients for stability guarantees:

- a generalized terminal cost term,
- a generalized terminal constraint, and
- a generalized controllability assumption.

This allows to consider both classes in a unified way and to find new stabilizing MPC schemes, which can be regarded as in-between of both classes. The potential advantages of these in-between schemes can be characterized as follows: In contrast to CLF-MPC, no control Lyapunov function is required, but only an upper bound on the optimal cost has to be known. In contrast to the previous results on unconstrained MPC, this upper bound does not have to hold globally, but is sufficient locally in a (possibly small) region around the origin, provided this region can be reached in finite time and an additional terminal constraint is added to the optimization problem.

3.5.1 MPC Setup with Integral Terminal Cost

The open-loop finite horizon optimal control problem at sampling instant t_i given measured state $x(t_i)$ is formulated as follows.

Problem 3.38.

$$\underset{\bar{u}\in\mathcal{PC}([t_i,t_i+T],\mathbb{R}^m)}{\text{minimize}} \quad J_{T,\tau}(x(t_i),\bar{u}) \tag{3.58a}$$

subject to

$$\dot{\bar{x}}(t';x(t_i),t_i) = f(\bar{x}(t';x(t_i),t_i),\bar{u}(t')) , \qquad t' \in [t_i,t_i+T] , \tag{3.58b}$$

$$\bar{x}(t_i;x(t_i),t_i) = x(t_i) , \tag{3.58c}$$

$$\bar{u}(t') \in \mathbb{U} , \qquad t' \in [t_i,t_i+T] , \tag{3.58d}$$

$$\bar{x}(t'';x(t_i),t_i) \in \Omega , \qquad t'' \in [t_i+T-\tau,t_i+T] , \tag{3.58e}$$

in which

$$J_{T,\tau}(x(t_i),\bar{u}) = \int_{t_i}^{t_i+T} F(\bar{x}(t';x(t_i),t_i),\bar{u}(t'))\,dt' + \int_{t_i+T-\tau}^{t_i+T} G(\bar{x}(t';x(t_i),t_i),\bar{u}(t'))\,dt' , \tag{3.58f}$$

and $\tau \in (0,T]$ is a constant design parameter.

We use an integral term, defined on an interval of length τ as terminal cost functional in the definition of the finite horizon cost functional $J_{T,\tau}$. This is in contrast to the use of only one single terminal cost term evaluated at the end of the prediction horizon T as common in the literature. Consequently, we do not only impose the terminal constraint on the "last" state, i.e., $\bar{x}(t_i+T;x(t_i),t_i) \in \Omega$, but actually require all states after a certain time $t_i+T-\tau$ to lie within Ω. The design parameter τ gives us flexibility in our MPC setup and allows to recover both, unconstrained MPC results and CLF-MPC. For more details on connections to previous MPC schemes, we refer to Section 3.5.3.

For a given sampling time δ with $0 < \delta \leq T$, the control input to the system is again defined in the usual continuous-time receding horizon fashion.

Algorithm 3.39 (Model Predictive Control with Integral Terminal Cost). *At each sampling instant $t_i = i\delta$, $i \in \mathbb{N}_0$, measure the state $x(t_i)$ and solve Problem 3.38. Apply the input*

$$u_{\mathrm{MPC}}(t) = u_T^\star(t;x(t_i),t_i) , \quad t_i \leq t < t_i+\delta \tag{3.59}$$

to the system until the next sampling instant $t_{i+1} = t_i + \delta$.

3.5.2 Asymptotic Stability

We make the following assumptions on the weighting functions F and G and on the terminal region Ω.

Assumption 3.12. *The stage cost $F : \mathbb{R}^n \times \mathbb{U} \to \mathbb{R}_{\geq 0}$ and the terminal cost function $G : \mathbb{R}^n \times \mathbb{U} \to \mathbb{R}_{\geq 0}$ are continuous, $F(0,0) = G(0,0) = 0$ and there are class $\mathcal{K}_\infty$ functions $\underline{\alpha}_F, \underline{\alpha}_G : \mathbb{R}_{\geq 0} \to \mathbb{R}_{\geq 0}$ such that for all $x \in \mathbb{R}^n$ and $u \in \mathbb{U}$*

$$F(x,u) \geq F(x,0) \geq \underline{\alpha}_F(|x|) \tag{3.60a}$$

$$and \quad G(x,u) \geq G(x,0) \geq \underline{\alpha}_G(|x|) . \tag{3.60b}$$

The terminal region $\Omega \subseteq \mathbb{R}^n$ is a closed set and contains $0 \in \mathbb{R}^n$ in its interior.

In order to simplify the subsequent analysis, we also make the following technical assumption throughout the remainder of this section.

Assumption 3.13. *The sampling time δ and the prediction horizon T are chosen such that*

$$\delta < T/2 \,.$$

In most practical cases, a small sampling time is desirable in any case due to robustness considerations. Moreover, similar results to the ones presented in this section can also be derived for $\delta \geq T/2$ as well, but are omitted for a concise presentation.

In the following analysis, similar to the previous sections, we consider the two consecutive sampling instants $t_0 = 0$ and $t_1 = \delta$ without loss of generality. Since system (3.1) is time-invariant, all results hold analogously for any other two consecutive sampling instants t_i and t_{i+1}. With slight abuse of notation, we use the following abbreviations

$$F^*(t; t_i) = F(x^*(t; x(t_i), t_i), u_T^*(t; x(t_i), t_i)) \,, \tag{3.61a}$$

$$G^*(t; t_i) = G(x^*(t; x(t_i), t_i), u_T^*(t; x(t_i), t_i)) \,, \tag{3.61b}$$

for $t \in [t_i, t_i + T]$ and $t_i \in \{0, \delta\}$. From this definition, it directly follows that

$$J_{T,\tau}^*(x(t_i)) = \int_{t_i}^{t_i+T} F^*(t'; t_i) dt' + \int_{t_i+T-\tau}^{t_i+T} G^*(t'; t_i) dt' \,.$$

We will use the following general result on asymptotic stability to derive our main results. Also note the close relation to Proposition 3.9.

Lemma 3.40 (General Condition for Asymptotic Stability). *Suppose the relation*

$$J_{T,\tau}^*(x(\delta)) \leq \int_{\delta}^{T} F^*(t'; 0) dt' + \int_{T-\tau}^{T} G^*(t'; 0) dt' \,. \tag{3.62}$$

holds for all functions $F^*(\cdot; 0) : [0, T] \to \mathbb{R}_{\geq 0}$ and $G^*(\cdot; 0) : [0, T] \to \mathbb{R}_{\geq 0}$ resulting from the optimization problem 3.38. Then, the closed-loop using MPC is asymptotically stable.

Proof. Similar to the result in Chen (1997) and the proof of Theorem 3.16, it can be shown that $J_{T,\tau}^*(x)$ is continuous in x at the origin. Condition (3.62) directly guarantees that

$$J_{T,\tau}^*(x(\delta)) - J_{T,\tau}^*(x(0)) \leq - \int_{0}^{\delta} F^*(t'; 0) dt' \,.$$

Hence, $J_{T,\tau}^*$ is non-increasing along trajectories of the closed-loop and stability directly follows. Asymptotic stability follows from $\int_{0}^{\infty} F(x(t'), u(t')) dt' \leq J_{T,\tau}^*(x(0))$, the lower bound (3.60), and Barbalat's Lemma (Barbalat, 1959; Khalil, 2002). The uniform continuity of the integrand is hereby guaranteed because of the continuity of f and compactness of $\mathbb{U}$. $\qquad\square$

In order to guarantee asymptotic stability for our MPC scheme with integral terminal cost, we use the following generalized controllability assumption, which is closely related to (Grüne, 2009, Assumption 3.1), (Grüne et al., 2010a, Assumption 3.1) and Assumptions 3.5 and 3.9.

Assumption 3.14 (Generalized Controllability Assumption). *For all $T' \in [0, T]$ and $x_0 \in \Omega$, there exists a piece-wise continuous input trajectory $\hat{u}(\cdot; x_0)$ with $\hat{u}(t; x_0) \in \mathbb{U}$ for all $t \in [0, T']$ such that*

a) the corresponding state trajectory satisfies $\bar{x}_{\hat{u}}(t) \in \Omega$ for all $t \in [0, T']$ (positive invariance of Ω) and

b) the corresponding cost is bounded by

$$J_{T', \tau'}(x_0, \hat{u}) \le B_G(T') G(x_0, 0), \tag{3.63}$$

in which $B_G : \mathbb{R}_{\geq 0} \to \mathbb{R}_{\geq 0}$ is a continuous and positive function and $\tau' = \min\{T', \tau\}$.

The following lemma gives a direct consequence of the generalized controllability assumption.

Lemma 3.41 (Direct Consequence of Controllability Assumption 3.14). *Suppose that Assumptions 3.1–3.3, 3.12 and 3.13, and the Generalized Controllability Assumption 3.14 are satisfied for system (3.1) and let $\mathfrak{t} = \max\{\delta, T - \tau\}$. Then the following holds for all $t^* \in [\mathfrak{t}, T]$*

$$J_{T, \tau}^*(x(\delta)) \le \int_\delta^{t^*} F^*(t'; 0) dt' + \int_{T+\delta-\tau}^{\max\{T+\delta-\tau, t^*\}} G^*(t'; 0) dt' + B_G(T + \delta - t^*) G^*(t^*; 0), \tag{3.64a}$$

and the following holds for all $t^* \in [T - \tau, T]$

$$\int_{t^*}^{T} F^*(t'; 0) + G^*(t'; 0) dt' \le B_G(T - t^*) G^*(t^*; 0). \tag{3.64b}$$

Proof. For any $t^* \in [\mathfrak{t}, T]$ define the control trajectory $\tilde{u}_{t^*}$ by

$$\tilde{u}_{t^*}(t) = \begin{cases} u_T^*(t; x(0), 0), & t \in [\delta, t^*[\\ \hat{u}(t - t^*; x^*(t^*; x(0), 0)), & t \in [t^*, T + \delta] \end{cases}$$

in which $\hat{u}$ is the input trajectory from Assumption 3.14 for initial state $x^*(t^*; x(0), 0)$. Since $\tilde{u}_{t^*}$ is a feasible, but not necessarily optimal, solution to the finite horizon optimal control problem 3.38 for initial state $x(\delta)$, we obtain

$$J_{T, \tau}^*(x(\delta)) \le J_{T, \tau}(x(\delta), \tilde{u}_{t^*})$$

$$\le \int_\delta^{t^*} F^*(t'; 0) dt' + \int_{T+\delta-\tau}^{\max\{T+\delta-\tau, t^*\}} G^*(t'; 0) dt' + J_{T+\delta-t^*, \min\{T+\delta-t^*, \tau\}}(x^*(t^*; x(0), 0), \hat{u})$$

$$\stackrel{(3.63)}{\le} \int_\delta^{t^*} F^*(t'; 0) dt' + \int_{T+\delta-\tau}^{\max\{T+\delta-\tau, t^*\}} G^*(t'; 0) dt' + B_G(T + \delta - t^*) G^*(t^*; 0),$$

which proves (3.64a). In order to prove Inequality (3.64b), note that the principle of optimality guarantees optimality of $u_T^*(t; x(0), 0)$ when considering the end-piece of the optimal cost $J_{T,\tau}^*(x(0))$ on any interval $[t^*, T]$. For $t^* \in [T - \tau, T]$, we can use (3.63) and (3.64b) directly follows. □

Another intermediate result based on the Generalized Controllability Assumption can be stated as follows.

Lemma 3.42 (Calculation of γ_τ). *Suppose that Assumptions 3.1–3.3, 3.12, 3.13, and the Generalized Controllability Assumption 3.14 are satisfied for system (3.1) and $\tau > \delta$. Then,*

$$\int_{T+\delta-\tau}^{T} G^*(t'; 0)dt' \le \gamma_\tau \int_{T-\tau}^{T+\delta-\tau} G^*(t'; 0) \, dt' \,, \tag{3.65}$$

in which

$$\frac{1}{\gamma_\tau} = \exp\left(\int_{T-\tau}^{T+\delta-\tau} \frac{1}{B_G(T - t^*)} \, dt^*\right) - 1 \,. \tag{3.66}$$

Proof. Equation (3.64b) in Lemma 3.41 yields for all $t^* \in [T - \tau, T]$

$$\int_{t^*}^{T} G^*(t'; 0)dt' \le B_G(T - t^*) \, G^*(t^*; 0) \,. \tag{3.67}$$

Consider any piece-wise continuous function $G^*(t; 0) : [T - \tau, T] \to \mathbb{R}_{\ge 0}$ satisfying (3.67) and define the function $\widehat{G}^*(t) : [T - \tau, T + \delta - \tau] \to \mathbb{R}_{\ge 0}$ by

$$\widehat{G}^*(t) = \frac{\int_{T+\delta-\tau}^{T} G^*(t'; 0)dt'}{B_G(T - t)} e^{-\int_{T+\delta-\tau}^{t} \frac{1}{B_G(T - t')} dt'} \,. \tag{3.68}$$

In the next step, we show that for all piece-wise continuous functions $G^*(t; 0)$ satisfying (3.67), the following holds

$$\int_{T-\tau}^{T+\delta-\tau} \widehat{G}^*(t')dt' \le \int_{T-\tau}^{T+\delta-\tau} G^*(t'; 0)dt' \,. \tag{3.69}$$

To this end, note that the anti-derivative of $\widehat{G}^*$ is

$$\int^{t} \widehat{G}^*(t')dt' = -\int_{T+\delta-\tau}^{T} G^*(t'; 0)dt' e^{-\int_{T+\delta-\tau}^{t} \frac{1}{B_G(T - t')} dt'} + \mathfrak{C}$$

with $\mathfrak{C} \in \mathbb{R}$. Hence, $\widehat{G}^*$ satisfies (3.67) with equality for all $t^* \in [T - \tau, T + \delta - \tau]$ in the sense of

$$\int_{t^*}^{T+\delta-\tau} \widehat{G}^*(t')dt' + \int_{T+\delta-\tau}^{T} G^*(t'; 0)dt' = B_G(T - t^*) \, \widehat{G}^*(t^*) \,.$$

For any $G^*(t; 0)$ satisfying (3.67), we know that $G^*(\delta; 0) - \widehat{G}^*(\delta) \geq 0$ and

$$\int_{t^*}^{T+\delta-\tau} \left(G^*(t'; 0) - \widehat{G}^*(t') \right) dt' \leq B_G(T - t^*) \left(G^*(t^*; 0) - \widehat{G}^*(t^*) \right)$$

holds for all $t^* \in [T - \tau, T + \delta - \tau]$. Define $\mathfrak{G}(t^*) = G^*(T + \delta - \tau - t^*; 0) - \widehat{G}^*(T + \delta - \tau - t^*)$, for which $\mathfrak{G}(0) \geq 0$ and $\int_0^{t^*} \mathfrak{G}(t')dt' \leq B_G(\tau - \delta + t^*)\mathfrak{G}(t^*)$ for all $t^* \in [0, \delta]$. Due to the comparison lemma (Khalil, 2002), it follows that $\mathfrak{G}(t^*) \geq 0$. Consequently, Equation (3.69) holds. On the other hand, direct calculations show that

$$\int_{T-\tau}^{T+\delta-\tau} \widehat{G}^*(t') \, dt' = \frac{1}{\gamma_\tau} \int_{T+\delta-\tau}^{T} G^*(t'; 0)dt' \,, \tag{3.70}$$

with γ_τ defined by (3.66). Hence, using (3.69) and (3.70) implies (3.65). This completes the proof. $\qquad\square$

Using the previous intermediate results, we can finally state the main result of this section in the following theorem.

Theorem 3.43 (Stability of MPC with Integral Terminal Cost). *Suppose that Assumptions 3.1–3.3, 3.12, 3.13, and the Generalized Controllability Assumption 3.14 are satisfied for system* (3.1) *and*

$$\begin{array}{ll} \int_\delta^{\delta+\tau} \frac{1}{B_G(t')} dt' \geq 1 \,, & \text{if } \tau \leq \delta \\ (\Xi_\tau - 1)\gamma_\tau \leq 1 \,, & \text{if } \tau > \delta \end{array} \tag{3.71}$$

with γ_τ defined in Lemma 3.42 and

$$\frac{1}{\Xi_\tau} = 1 - \exp\left(-\int_{T+\delta-\tau}^{T} \frac{1}{B_G(T + \delta - t^*)} \, dt^* \right) \,. \tag{3.72}$$

Then, the closed-loop system resulting from the application of the model predictive controller according to Algorithm 3.39 to system (3.1) *is asymptotically stable. The region of attraction is the set of all initial conditions for which Problem 3.38 is initially feasible.*

Proof. We can distinguish three different cases A, B, and C for τ as shown in Figure 3.7, of which cases B and C can be treated in the same way if Assumption 3.13 holds.

Case A ($\tau \leq \delta$): Due to Lemma 3.41, Equation (3.64a), and $\mathbf{t} = T - \tau$, we obtain

$$J_{T,\tau}^*(x(\delta)) \leq \min_{t^* \in [T-\tau, T]} \left(\int_\delta^{t^*} F^*(t'; 0)dt' + B_G(T + \delta - t^*)G^*(t^*; 0) \right)$$

$$\leq \int_\delta^{T} F^*(t'; 0)dt' + \min_{t^* \in [T-\tau, T]} \left(B_G(T + \delta - t^*)G^*(t^*; 0) \right)$$

$$\leq \int_\delta^{T} F^*(t'; 0)dt' + \frac{1}{\int_{T-\tau}^{T} \frac{1}{B_G(T+\delta-t^*)}dt^*} \int_{T-\tau}^{T} G^*(t'; 0)dt' \,.$$

Hence, if

$$\frac{1}{\int\limits_{T-\tau}^{T} \frac{1}{B_G(T+\delta-t^*)} dt^*} \leq 1 \quad \text{or, equivalently,} \quad \int\limits_{\delta}^{\delta+\tau} \frac{1}{B_G(t')} dt' \geq 1, \quad (3.73)$$

then (3.62) holds and asymptotic stability follows from Lemma 3.40.

Cases B and C ($\tau > \delta$): We distinguish two subcases in the first step. In the first case, if $J_{T,\tau}^*(x(\delta)) \leq \int\limits_{\delta}^{T} F^*(t';0)dt'$, then (3.62) is satisfied and asymptotic stability follows directly from Lemma 3.40. Hence, we will only consider the second case $J_{T,\tau}^*(x(\delta)) > \int\limits_{\delta}^{T} F^*(t';0)dt'$ in the following. Equation (3.64a) in Lemma 3.41 and $F^* \geq 0$ then provide

$$J_{T,\tau}^*(x(\delta)) - \int\limits_{\delta}^{T} F^*(t';0)dt' \leq \int\limits_{T+\delta-\tau}^{t^*} G^*(t';0)dt' + B_G(T+\delta-t^*)G^*(t^*;0) \quad (3.74)$$

for all $t^* \in [T + \delta - \tau, T]$. Consider any piece-wise continuous function $G^*(t;0) : [T + \delta - \tau, T] \to \mathbb{R}_{\geq 0}$ satisfying (3.74) and define the function $\widehat{G}^*(t) : [T + \delta - \tau, T] \to \mathbb{R}_{\geq 0}$ by

$$\widehat{G}^*(t) = \frac{J_{T,\tau}^*(x(\delta)) - \int\limits_{\delta}^{T} F^*(t';0)dt'}{B_G(T+\delta-t)} e^{-\int\limits_{T+\delta-\tau}^{t} \frac{1}{B_G(T+\delta-t')} dt'}. \quad (3.75)$$

In the next step, we show that for all piece-wise continuous functions $G^*(t;0)$ satisfying (3.74), the following holds

$$\int\limits_{T+\delta-\tau}^{T} \widehat{G}^*(t')dt' \leq \int\limits_{T+\delta-\tau}^{T} G^*(t';0)dt'. \quad (3.76)$$

To this end, note that $\widehat{G}^*(t)$ satisfies (3.74) with equality (instead of inequality) for all $t^* \in [T+\delta-\tau, T]$, which can be shown by direct evaluation of (3.74) for $t^* = T+\delta-\tau$ and taking the derivative with respect to t^* on both sides of (3.74). This step is similar to the proof of Lemma 3.14. For the sake of contradiction, assume $\int\limits_{T+\delta-\tau}^{T} \widehat{G}^*(t')dt' > \int\limits_{T+\delta-\tau}^{T} G^*(t';0)dt'$. But then there exists a $t \in [T + \delta - \tau, T]$ for which

$$\int\limits_{T+\delta-\tau}^{t} \widehat{G}^*(t')dt' \geq \int\limits_{T+\delta-\tau}^{t} G^*(t';0)dt' \quad \text{and} \quad \widehat{G}^*(t) > G^*(t;0).$$

But this contradicts (3.74), which shows that (3.76) holds. On the other hand, direct calculations reveal

$$J_{T,\tau}^*(x(\delta)) - \int\limits_{\delta}^{T} F^*(t';0)dt' = \Xi_\tau \int\limits_{T+\delta-\tau}^{T} \widehat{G}^*(t')dt', \quad (3.77)$$

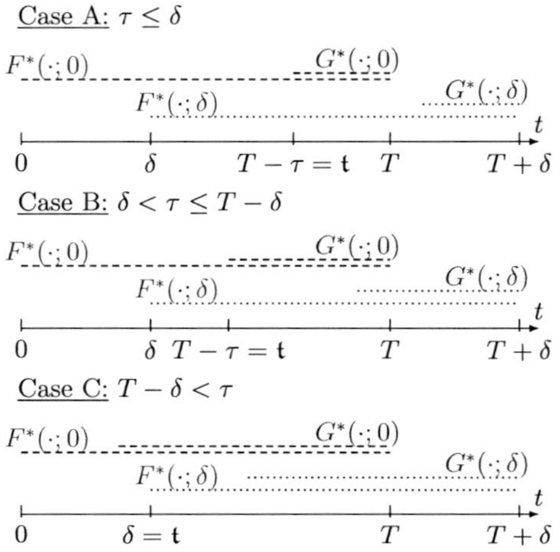

Figure 3.7: Sketch of the Generalized Terminal Cost and possible different cases depending on τ.

in which Ξ_τ is defined in (3.72). Combining (3.76) and (3.77) yields

$$J_{T,\tau}^*(x(\delta)) - \int_\delta^T F^*(t';0)dt' \leq \Xi_\tau \int_{T+\delta-\tau}^T G^*(t';0)dt'$$

$$\overset{(3.65)}{\leq} (\Xi_\tau - 1)\,\gamma_\tau \int_{T-\tau}^{T+\delta-\tau} G^*(t';0)dt' + \int_{T+\delta-\tau}^T G^*(t';0)dt'.$$

Thus, if $(\Xi_\tau - 1)\,\gamma_\tau \leq 1$, then Inequality (3.62) holds and asymptotic stability follows from Lemma 3.40. $\qquad\square$

3.5.3 Connection to Previous Results and Special Cases

In this section, we discuss the connections of the proposed unifying scheme to previous MPC schemes, namely CLF-MPC and unconstrained MPC.

CLF-MPC

In this paragraph, we investigate the connections of our proposed scheme to the classical results on CLF-MPC, see, e.g., Chen and Allgöwer (1998); Fontes (2001); Mayne et al. (2000) and Section 2.1.3.

Following the references cited, we assume the existence of a terminal cost function $E : \mathbb{R}^n \to \mathbb{R}_{\geq 0}$ satisfying the following assumptions.

Assumption 3.15. *E is continuously differentiable, $E(0) = 0$, and there is class $\mathcal{K}_\infty$ function $\underline{\alpha}_E : \mathbb{R}_{\geq 0} \to \mathbb{R}_{\geq 0}$ such that for all $x \in \mathbb{R}^n$*

$$E(x) \geq \underline{\alpha}_E(|x|).$$

Furthermore, the terminal region $\Omega \subseteq \mathbb{R}^n$ is a closed set, contains $0 \in \mathbb{R}^n$ in its interior, and there is a local feedback law $k : \Omega \to \mathbb{U}$ such that for all $x \in \Omega$

$$\frac{\partial E}{\partial x} f(x, k(x)) \leq -F(x, k(x)) - \varepsilon E(x). \tag{3.78}$$

We can choose $G(x, u) = \frac{1}{\tau} E(x)$ in our setup with integral terminal cost in order to approximate the terminal cost function E. The following stability result is a consequence of our main result stated in Theorem 3.43.

Proposition 3.44. *Suppose that Assumptions 3.1–3.3, 3.12, 3.13, and 3.15 are satisfied for system (3.1) and $G(x, u) = \frac{1}{\tau} E(x)$. Then, for all $\varepsilon > 0$, there exists a $\tau > 0$ sufficiently small such that the closed-loop system resulting from the application of the model predictive controller according to Algorithm 3.39 to system (3.1) is asymptotically stable.*

Proof. Since we only need to find a τ sufficiently small, we can assume $\tau < \delta$ in the following. Consider any $x_0 \in \Omega$ and denote the closed-loop trajectory resulting from the local feedback k by $\bar{x}(t; x_0)$ for $t \in \mathbb{R}_{\geq 0}$, i.e.,

$$\dot{\bar{x}}(t; x_0) = f(\bar{x}(t; x_0), k(\bar{x}(t; x_0))), \quad \bar{x}(0; x_0) = x_0.$$

Furthermore, define the following abbreviations

$$\hat{u}(t; x_0) = k(\bar{x}(t; x_0)), \quad \bar{F}_k(t) = F(\bar{x}(t; x_0), k(\bar{x}(t; x_0))), \quad \text{and} \quad \bar{E}_k(t) = E(\bar{x}(t; x_0)).$$

Due to (3.78), the following holds for all $T \in \mathbb{R}_{\geq 0}$

$$\int_0^T \left(\bar{F}_k(t') + \varepsilon \bar{E}_k(t') \right) dt' + \bar{E}_k(T) \leq \bar{E}_k(0),$$

and consequently

$$
\begin{aligned}
J_{T,\tau}(x_0, \hat{u}) &= \int_0^T \bar{F}_k(t') dt' + \int_{T-\tau}^T \frac{\bar{E}_k(t')}{\tau} dt' = \frac{1}{\tau} \int_{T-\tau}^T \left(\int_0^T \bar{F}_k(t'') dt'' + \bar{E}_k(t') \right) dt' \\
&\leq \frac{1}{\tau} \int_{T-\tau}^T \left(\bar{E}_k(0) - \bar{E}_k(T) - \varepsilon \int_0^{t'} \bar{E}_k(t'') dt'' + \bar{E}_k(t') \right) dt' \\
&= \bar{E}_k(0) + \frac{1}{\tau} \int_{T-\tau}^T \left(\bar{E}_k(t') - \bar{E}_k(T) - \varepsilon \int_0^{t'} \bar{E}_k(t'') dt'' \right) dt'.
\end{aligned}
$$

Moreover, we know that $\bar{E}_k(t) \leq e^{-\varepsilon(t-(T-\tau))} \bar{E}_k(T - \tau)$ for $t \geq T - \tau$ in account of (3.78). Hence, $\int_{T-\tau}^T \bar{E}_k(t') dt' \leq \frac{1-e^{-\varepsilon\tau}}{\varepsilon} \bar{E}_k(T - \tau)$. On the other hand, $\int_{T-\tau}^T \varepsilon \int_0^{t'} \bar{E}_k(t'') dt'' dt' \geq \varepsilon(T -$

$\tau)\bar{E}_k(T-\tau)$. If $\tau > 0$ is chosen small enough such that $1 - e^{-\varepsilon\tau} \leq \varepsilon^2(T-\tau)$ or, equivalently, $\tau \leq -\ln(1 - \varepsilon^2(T-\tau))/\varepsilon$, then

$$J_{T,\tau}(x_0, \hat{u}) \leq \bar{E}_k(0) = \tau\, G(x_0, 0)\,.$$

Therefore, Assumption 3.14 is satisfied with $B_G(T) = \tau$ and the result follows from Theorem 3.43. $\qquad\square$

Remark 3.45. *In the limit case $\tau \to 0$, ε in (3.78) can be chosen arbitrarily close to zero and the integral terminal cost term $\int_{T-\tau}^{T} G(x, u)dt'$ becomes $E(x(T))$. Hence, Proposition 3.44 recovers the stability result of the classical CLF-MPC, see Theorem 2.6 and (Chen and Allgöwer, 1998; Fontes, 2001; Mayne et al., 2000). It is evident that the proof of this result via the limit case of an integral terminal cost term is more complicated and less attractive than the direct proofs given in the references (Chen and Allgöwer, 1998; Fontes, 2001; Mayne et al., 2000; Rawlings and Mayne, 2009). However, it demonstrates the connection between our proposed framework and classical results.*

Unconstrained MPC

In this paragraph, we discuss the connection of our proposed scheme to existing unconstrained MPC schemes, see, e.g., the results in discrete-time by (Grimm et al., 2005; Grüne, 2009; Grüne et al., 2010a) and the results for continuous-time systems in Section 3.2.

The choice of $\tau = T$, $\Omega = \mathbb{R}^n$, and $G(x, u) = \lambda F(x, u)$ with some constant $\lambda \in \mathbb{R}_{>0}$ directly recovers the unconstrained MPC setup without additional terminal cost terms. Furthermore, Assumption 3.14 is satisfied for $B_G(t) = \frac{1+\lambda}{\lambda} B(t)$, in which B was defined in Assumption 3.5. Comparing the results of Theorem 3.43 to the main result of Section 3.2 stated in Theorem 3.16 reveals that we recover the stability results of the unconstrained MPC scheme for $\lambda \to \infty$.

It can be shown that this estimate is optimal in the sense of being a solution to an infinite-dimensional optimization problem based on the controllability assumption, see Section 3.2 and, in particular, Theorem 3.18 for more details.

3.5.4 Improved Stability Conditions

As discussed in Section 3.5.3, the conditions for asymptotic stability presented in Theorem 3.43 do not fully recover the results of Theorem 3.16 if a finite λ is chosen. Indeed, the conditions are more conservative in some sense. One reason for the conservativeness is that we have not made any assumptions so far about a connection between F and G in Section 3.5, respectively. Hence, we will use the following assumption in the following.

Assumption 3.16 (Compatibility of F and G). *There exist positive constants $0 < \underline{\lambda} \leq \overline{\lambda}$ such that for all $x \in \mathbb{R}^n$, $u \in \mathbb{R}^m$*

$$\underline{\lambda}\, F(x, u) \leq G(x, u) \leq \overline{\lambda}\, F(x, u)\,. \tag{3.79}$$

For the investigation of an unconstrained MPC setup as in Theorem 3.16, it is clear that an even stronger condition holds, namely $G(x, u) = \lambda F(x, u)$ with $\underline{\lambda} = \overline{\lambda} = \lambda$. Furthermore, Assumption 3.16 is always satisfied when considering quadratic cost terms.

Close inspection reveals that Cases A and B do not change when using Assumption 3.16 in the proof of Theorem 3.43. Indeed, Case C is the interesting part with the connection to (Reble and Allgöwer, 2011, Theorem 6). Combining (3.64b) and (3.79) yields for all $t^* \in [T - \tau, T]$

$$\int_{t^*}^{T} G^*(t'; 0) dt' \leq B_{\overline{\lambda}}(T - t^*) G^*(t^*; 0), \qquad (3.80)$$

in which $B_{\overline{\lambda}}(T') = \frac{\overline{\lambda}}{\overline{\lambda}+1} B_G(T')$. Careful inspection reveals that the result of Lemma 3.42 then holds with $B(t)$ replaced by $B_{\overline{\lambda}}(t)$ in (3.66). Since $B_{\overline{\lambda}}(t) \leq B(t)$, this results in better estimates of the prediction horizon required to guarantee asymptotic stability.

These results in combination with $B_G(t) = \frac{1+\lambda}{\lambda} B(t)$ as discussed in Section 3.5.3 allow to recover the results on unconstrained MPC of Theorem 3.16.

3.5.5 Illustrative Example

Consider the unstable nonlinear system

$$\dot{x}_1(t) = x_1(t) + u(t), \qquad (3.81a)$$
$$\dot{x}_2(t) = (x_1(t) + u(t))^3, \qquad (3.81b)$$

with input constraint $\mathbb{U} = [-3, 3]$. By a simple input transformation, the system can be transformed into the cubic integrator, which was considered in Section 3.3.3 and in Grimm et al. (2005) formulated in a discrete-time version. Similar to the cubic integrator, the system is not stabilizable by continuous state feedback and, in particular, the Jacobi linearization of the system is not asymptotically stabilizable. Therefore, the design of a corresponding control Lyapunov function is a difficult task, which makes CLF-MPC with any non-trivial terminal region $\Omega \neq \{0\}$ unattractive for this problem. Although it is difficult to design a stabilizing state feedback for this problem, an open-loop control is much more simple to obtain. To this end, we consider the input transformation $v(t) = u(t) + x_1(t)$ and the open-loop control given by four piece-wise constant transformed input values $v(t) = v_i$ for $t \in [i - 1, i[$ and $i \in \{1, \ldots, 4\}$. Choosing v_i as

$$\left\{ -x_1(0), \left(-0.5 + \sqrt{7/12} \right) \Psi, \left(-0.5 - \sqrt{7/12} \right) \Psi, \Psi \right\},$$

in which $\Psi = (x_2(0) - x_1(0)^3)^{1/3}$, drives the system to the origin at time $t = 4$, see Section 3.3.3 and Grimm et al. (2005). Due to the input constraints, it is clear that this open-loop control is only feasible in some region around the origin. Hence, we will consider only the region

$$\Omega = \{x \,:\, x_1 \in [-1, 1], x_2 \in [-1 + x_1^3, 1 + x_1^3]\} \qquad (3.82)$$

for our controllability assumption. Figure 3.8 shows a sketch of Ω and note that $|\Psi| \leq 1$ for all $x(0) \in \Omega$, which guarantees that the input constraints are satisfied for the open-loop control given above. One can show that Ω is invariant under the open-loop control defined

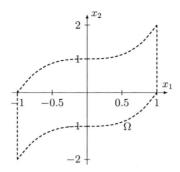

Figure 3.8: Sketch of the terminal region Ω defined in (3.82) for system (3.81).

above. For example, calculation of the state trajectory generated by the open-loop input yields

$$x(1) = \begin{pmatrix} 0 \\ \Psi^3 \end{pmatrix}, \ x(2) = \begin{pmatrix} \left(-0.5 + \sqrt{7/12}\right)\Psi \\ \left(1 + (-0.5 + \sqrt{7/12})^3\right)\Psi^3 \end{pmatrix}, \ x(3) = \begin{pmatrix} -\Psi \\ -\Psi^3 \end{pmatrix}, \ x(4) = 0.$$

If we choose $F(x, u) = G(x, u) = x_1^6 + x_2^2 + u^6$, straightforward calculations show for all $T \geq \tau > 0$ and all $x_0 \in \Omega$

$$J_{T,\tau}^*(x_0) \leq 144\, G(x_0, 0).$$

Asymptotic stability is then guaranteed by Theorem 3.43 for $\tau > 718$ and $\delta = 1$. The estimate of the minimal prediction horizon can be improved by using $u(t) = -x_1(0)$ as another simple choice of an open-loop input, which shows $J_{T,\tau}^*(x_0) \leq 2(T + \tau)\, G(x_0, 0)$ and guarantees asymptotic stability for $\tau > 484$.

The region of attraction contains all initial conditions, for which the system can be steered to Ω in finite time $T - \tau$. In contrast, an "unconstrained" MPC scheme without additional integral terminal cost ($G(x, u) = 0$) would only guarantee asymptotic stability for all initial conditions $x_0 \in \Omega$ and would also require invariance of Ω as an additional constraint. This underpins the advantages of our MPC scheme with integral terminal cost.

Note that the results on the prediction horizon are still conservative and can be further improved by using additional information.

3.5.6 Summary

In this section, we presented a first step towards a unifying view on unconstrained MPC schemes and MPC schemes using terminal constraints and control Lyapunov functions as terminal weight. A novel MPC scheme using an integral terminal cost term has been proposed and conditions for asymptotic stability have been derived. Both classes of previous MPC schemes can be obtained as limit cases of this proposed framework. Furthermore, the possible advantages of our novel MPC scheme were illustrated in a brief example.

3.6 MPC with Exponential Weighting and Terminal Constraints

Besides guaranteeing asymptotic stability for a possibly shorter prediction horizon, the exponential weighting introduced in Section 3.4 has another advantage: If the exponential controllability assumption is only satisfied locally in a (terminal) region Ω around the origin and appropriately defined terminal constraints are added to the finite horizon optimal control problem, we are still able to show stability of the closed-loop. Note that this is not possible for the MPC schemes proposed in Sections 3.2 and 3.3 unless $\Gamma = 1$, see also Remark 3.49 after Theorem 3.48 in Section 3.6.2. However, a similar result was presented in Section 3.5 by considering a generalized integral cost. In this section, the exponential weighting provides a similar additional weighting term as the generalized integral cost in Section 3.5.

3.6.1 MPC Setup with Exponential Weighting and Terminal Constraints

In this section, we consider a terminal region $\Omega \subset \mathbb{R}^n$ which is closed and contains the origin in its interior. The open-loop finite horizon optimal control problem at sampling instant t_i given measured state $x(t_i)$ is formulated as follows.

Problem 3.46.

$$\underset{\bar{u}\in\mathcal{PC}([t_i,t_i+T],\mathbb{R}^m)}{\text{minimize}} \quad J_{T,\mu,\tau}(x(t_i),\bar{u}) \tag{3.83a}$$

subject to

$$\dot{\bar{x}}(t';x(t_i),t_i) = f(\bar{x}(t';x(t_i),t_i),\bar{u}(t')), \qquad t' \in [t_i, t_i+T], \tag{3.83b}$$

$$\bar{x}(t_i;x(t_i),t_i) = x(t_i), \tag{3.83c}$$

$$\bar{u}(t') \in \mathbb{U}, \qquad t' \in [t_i, t_i+T], \tag{3.83d}$$

$$\bar{x}(t'';x(t_i),t_i) \in \Omega, \qquad t'' \in [t_i+T-\tau, t_i+T], \tag{3.83e}$$

in which

$$J_{T,\mu,\tau}(x(t_i),\bar{u}) = \int_{t_i}^{t_i+T} \beta(t'-t_i)\, F(\bar{x}(t';x(t_i),t_i),\bar{u}(t'))\, dt' \tag{3.83f}$$

in which $\beta(t) = e^{\mu t}$ for some constant $\mu \in \mathbb{R}_{\geq 0}$ is an exponential weighting on the stage cost and $\tau \in (0, T-\delta]$ is a constant design parameter.

Note that we do not only impose the terminal constraint on the "last" state, i.e., $\bar{x}(t_i + T; x(t_i), t_i) \in \Omega$, but actually require all states after a certain time $t_i + T - \tau$ to lie within Ω, which is similar to the generalized terminal constraint considered in Section 3.5.

For a given sampling time δ with $0 < \delta \leq T$, the control input to the system is again defined in the usual continuous-time receding horizon fashion.

Algorithm 3.47 (Model Predictive Control with Local Exponential Controllability). *At each sampling instant $t_i = i\delta$, $i \in \mathbb{N}_0$, measure the state $x(t_i)$ and solve Problem 3.46. Apply the input*

$$u_{\mathrm{MPC}}(t) = u_T^*(t; x(t_i), t_i), \quad t_i \le t < t_i + \delta \tag{3.84}$$

to the system until the next sampling instant $t_{i+1} = t_i + \delta$.

3.6.2 Asymptotic Stability

First, we make the following standard assumption regarding the terminal region.

Assumption 3.17. *The terminal region Ω is a closed set and contains $0 \in \mathbb{R}^n$ in its interior.*

Additionally, we modify Assumption 3.11 as follows.

Assumption 3.18 (Local Exponential Controllability). *For all $x_0 \in \Omega$, there exists a piece-wise continuous input trajectory $\hat{u}(\cdot; x_0, 0)$ with $\hat{u}(t; x_0, 0) \in \mathbb{U}$ for all $t \in \mathbb{R}_{\ge 0}$ such that the corresponding state trajectory $\bar{x}_{\hat{u}}$ satisfies for all $t \in \mathbb{R}_{\ge 0}$*

$$F(\bar{x}_{\hat{u}}(t; x_0, 0), \hat{u}(t; x_0, 0)) \le C\, e^{-\lambda t}\, F(x_0, 0) \qquad and \qquad \bar{x}_{\hat{u}}(t; x_0, 0) \in \Omega\,,$$

with overshoot constant $C \ge 1$ and decay rate $\lambda > \mu > 0$.

There are two main differences compared to Assumption 3.11. On the one hand, we now only require the exponential controllability assumption to be satisfied locally in a (terminal) region Ω around the origin. On the other hand, we additionally assume that this region is controlled positively invariant.

We can summarize the main result of this section as follows.

Theorem 3.48 (Stability of MPC with Exponential Weighting and Terminal Constraints). *Suppose that Assumptions 3.1–3.4, 3.17, and Exponential Controllability Assumption 3.18 are satisfied for the system (3.1). Furthermore, assume that $\Xi_{\mu,\tau} \le 1$ for $\Xi_{\mu,\tau}$ defined by*

$$\frac{e^{-\mu\delta}}{\Xi_{\mu,\tau}} = 1 - \left(\frac{e^{(\lambda-\mu)\delta} - 1}{e^{(\lambda-\mu)\tau} - 1} \right)^{\frac{1}{C}}. \tag{3.85}$$

Then, the closed-loop system resulting from the application of the model predictive controller according to Algorithm 3.47 to system (3.1) is asymptotically stable. The region of attraction is the set of all initial conditions for which Problem 3.46 is initially feasible.

Proof. We can distinguish two cases:

$$(i)\ J_{T,\mu,\tau}^*(x(\delta)) \le \int_\delta^T \beta(t')F^*(t';0)dt' \quad \text{and} \quad (ii)\ J_{T,\mu,\tau}^*(x(\delta)) > \int_\delta^T \beta(t')F^*(t';0)dt'\,.$$

For case (i), Inequality (3.86) below is directly satisfied. For case (ii), we can follow the proof of Lemma 3.35. Due to the definitions made and the local exponential controllability assumption, we can show that (3.55) holds for all $t^* \in [T - \tau, T]$. Hence, $\int_\delta^T \beta(t')\widehat{F}^*(t')dt' \leq \int_\delta^T \beta(t')F^*(t';0)dt'$ with $\widehat{F}^*(t) : [\delta, T] \to \mathbb{R}_{\geq 0}$ defined by

$$\widehat{F}^*(t) = \begin{cases} F^*(t';0), & \delta \leq t < T - \tau, \\ e^{-\mu(t-\delta)} \dfrac{J_{T,\mu,\tau}^*(x(\delta)) - \int_\delta^{T-\tau} \beta(t')\widehat{F}^*(t')dt'}{B_\mu(T+\delta-t)} e^{-\int_{T-\tau}^t \frac{1}{B_\mu(T+\delta-t')} dt'}, & T - \tau \leq t \leq T. \end{cases}$$

Direct calculations show that

$$J_{T,\mu,\tau}^*(x(\delta)) = \int_\delta^{T-\tau} \beta(t')\widehat{F}^*(t')dt' + \Xi_{\mu,\tau} \int_\delta^T \beta(t')\widehat{F}^*(t')dt'.$$

Thus, similar to (3.53a) and (3.54a), we obtain

$$J_{T,\mu,\tau}^*(x(\delta)) \leq \max\{\Xi_{\mu,\tau}, 1\} \cdot \int_\delta^T \beta(t')F^*(t';0)dt'. \tag{3.86}$$

Hence, Inequality (3.86) is satisfied for both cases (i) and (ii). For $\Xi_{\mu,\tau} \leq 1$, it directly follows that

$$J_{T,\mu,\tau}^*(x(\delta)) \leq J_{T,\mu,\tau}^*(x(0)) - \int_0^\delta \beta(t')F^*(t';0)dt'$$

and, consequently, asymptotic stability is guaranteed by analogue arguments to the proof of Theorem 3.16 when replacing J_T^* by $J_{T,\mu,\tau}^*$. $\qquad\square$

Remark 3.49. *The stability guarantee relies on the fact that $\mu > 0$. Using only a local controllability assumption does not allow to conclude (3.40) in Lemma 3.27 or (3.53b). Hence, in order to show that $\alpha_\mu > 0$ in (3.57), we require $\Xi_E \leq 1$ in (3.34) or $\Xi_{\mu,\tau} \leq 1$ in (3.53a), respectively. However, this is not possible for any finite horizon unless an additional weighting is employed.*

The result in this section can be regarded as in-between those established in (Chen and Allgöwer, 1998; Fontes, 2001; Mayne et al., 2000) for MPC schemes using a terminal constraint and a terminal cost, and the stability results for unconstrained MPC schemes based on a controllability assumption, see (Grimm et al., 2005; Grüne, 2009; Grüne and Pannek, 2011; Grüne et al., 2010a) and Sections 3.2 and 3.3. The advantages of this in-between scheme can be summarized as follows: In contrast to (Chen and Allgöwer, 1998; Fontes, 2001; Mayne et al., 2000), no control Lyapunov function is required. In contrast to (Grimm et al., 2005; Grüne, 2009; Grüne and Pannek, 2011; Grüne et al., 2010a; Reble and Allgöwer, 2012b), the controllability assumption does not have to be satisfied globally, but is only required locally in a (possibly small) region around the origin. A possible drawback of the current approach is that the region Ω has to be reachable in finite time and an additional terminal constraint is added to the optimization problem. The latter does not only confine the terminal state at the end of the prediction horizon, but all predicted states in some interval of length τ.

3.6.3 Summary

In this section, we have shown that an additional exponential weighting on the stage cost allows to guarantee stability when only using a local controllability assumption in combination with appropriate terminal constraints. This result is similar to the results when using an integral terminal cost as introduced in Section 3.5.

3.7 Summary

In this chapter, we proposed five novel MPC schemes with explicit conditions on the length of the prediction horizon in order to guarantee asymptotic stability of the closed-loop, see also Figure 3.1 for a schematic overview. The cornerstone of the stability analysis of all five schemes is an asymptotic controllability assumption, which requires the knowledge of an appropriate upper bound on the optimal cost function in terms of the stage cost. Hence, this assumption is less restrictive than the knowledge of a control Lyapunov function as classically assumed in order to guarantee stability with MPC.

The main properties of the different schemes are summarized in Table 3.1. As already mentioned, all five schemes allow stability guarantees without the knowledge of a local control Lyapunov function. However, the schemes with exponential weighting proposed in Sections 3.4 and 3.6 require the more restrictive exponential controllability assumption in contrast to the asymptotic controllability assumption sufficient for the other approaches. The three schemes proposed in Sections 3.2–3.4 each rely on a global controllability assumption, more precisely a global upper bound on the optimal cost. In contrast, local information is sufficient for the schemes in Sections 3.5 and 3.6, albeit additional terminal constraints have to be added to the optimal control problem. Finally, the simplest MPC setup, i.e., MPC without terminal cost and without terminal constraints, has one advantage compared to all other setups in providing a guaranteed performance estimate of the closed-loop.

The development of these novel MPC schemes also brings several new possible future research directions, for which we refer to Section 5.2 at the end of this thesis.

Table 3.1: Comparison of the different MPC schemes for nonlinear continuous-time systems considered in Chapter 3.

Design scheme	Properties	no local CLF required	only asymptotic controllability required	no terminal constraints	only local controllability assumption	suboptimality estimate
Unconstrained MPC (Section 3.2)		✓	✓	✓	–	✓
Unconstrained MPC with general terminal cost (Section 3.3)		✓/–	✓	✓	–	–
Unconstrained MPC with exponential weighting (Section 3.4)		✓	–	✓	–	–
MPC with integral terminal cost (Section 3.5)		✓	✓	–	✓	–
MPC with exponential weighting and terminal constraints (Section 3.6)		✓	–	–	✓	–

Chapter 4

Model Predictive Control for Nonlinear Time-Delay Systems

As discussed in Section 2.2.3, several methods for the stabilizing control of nonlinear time-delay systems have been proposed in the literature, but most of these methods do not allow to take hard constraints into account. For systems with constraints, model predictive control (MPC) is an attractive choice as control method in general. While there exists a significant number of publications on MPC for systems without delays in the states, see Section 2.1.2, only few results are available concerning MPC for nonlinear time-delay systems. In these results, stability of the closed-loop is guaranteed either by using a global control Lyapunov functional (CLF) as terminal cost (Kwon et al., 2001a,b; Lu, 2011; Mahboobi Esfanjani and Nikravesh, 2009a) or by an extended zero terminal state constraint (Angrick, 2007; Raff et al., 2007). Both approaches are less attractive for different reasons. The construction of a global CLF or a globally stabilizing controller is particularly difficult in the presence of input constraints and can only be expected to be feasible in rare special cases. Furthermore, the extended zero terminal state constraint is particularly problematic from a computational point of view. The exact satisfaction of the constraint does require an infinite number of iterations in the numerical optimization and feasibility problems may occur for short prediction horizons. To overcome these difficulties, we investigate the use of alternative MPC schemes for nonlinear time-delay systems in this chapter. For a schematic overview of these schemes, see Figure 4.1.

First, we extend the well-known stability results for MPC with terminal cost and terminal constraints, see Section 2.1.3 and the references therein, to nonlinear time-delay systems. Our results contain the previous results for time-delay systems (using a global CLF or an extended zero terminal state constraint) as special cases and are very similar to the results for finite-dimensional systems. The main difference is the use of an appropriate terminal cost functional instead of a terminal cost function and minor additional technical details in the proof. Due to the infinite-dimensional nature of nonlinear time-delay systems, more significant difficulties are encountered for the calculation of the terminal cost and terminal constraints based on the Jacobi linearization following the well-known procedure presented by Chen and Allgöwer (1998). We propose four different procedures in order to overcome these difficulties. Although each procedure has different additional assumptions and properties, each one contains the results for systems without delays as special case.

Second, two complementary unconstrained MPC schemes for nonlinear time-delay systems are presented. In the first scheme, a local control Lyapunov functional is employed as in the MPC scheme with terminal constraints. Despite removing the terminal constraint from the optimal control problem, its satisfaction is nevertheless guaranteed for a defined set of initial states, thereby extending previous results for finite-dimensional systems presented by Limon et al. (2006). In the second scheme, we extend the results of Section 3.2 and show stability without terminal cost terms. A stabilizing minimal prediction horizon is calculated based on a controllability assumption.

The remainder of this chapter is organized as follows. We present the problem setup considered in this chapter in Section 4.1. In Section 4.2, we give stability conditions for MPC with terminal cost and terminal constraints for nonlinear time-delay systems. In Section 4.3, we propose four procedures for calculating the terminal cost and terminal constraints based on the Jacobi linearization and discuss their properties. In Sections 4.4 and 4.5, we derive stability conditions for unconstrained MPC with and without terminal cost, respectively. In Section 4.6, we illustrate and compare the results of this chapter with two numerical examples: an academic example and a continuous stirred tank reactor with recycle stream. Last, the results of this chapter are summarized in Section 4.7.

Parts of this chapter are based on Mahboobi Esfanjani et al. (2009); Reble and Allgöwer (2010a,b, 2012a); Reble et al. (2011a,b,c).

4.1 Problem Setup

In this chapter, we consider nonlinear time-delay systems in continuous-time described by the functional differential equation (FDE)

$$\dot{x}(t) = f(x(t), x(t-\tau), u(t)), \tag{4.1a}$$
$$x(\theta) = \varphi(\theta), \qquad \forall \theta \in [-\tau, 0], \tag{4.1b}$$

in which $x(t) \in \mathbb{R}^n$ is the instantaneous state at time t, $x(t-\tau) \in \mathbb{R}^n$ is the delayed state, and $u(t) \in \mathbb{R}^m$ is the control input subject to input constraints $u(t) \in \mathbb{U} \subset \mathbb{R}^m$. The time-delay $\tau \in \mathbb{R}_{>0}$ is constant and assumed to be known. The initial function is given by $\varphi \in \mathcal{C}_\tau$, in which $\mathcal{C}_\tau = \mathcal{C}([-\tau, 0], \mathbb{R}^n)$ denotes the Banach space of continuous functions mapping the interval $[-\tau, 0] \subset \mathbb{R}$ into $\mathbb{R}^n$.

We will use the following assumptions throughout this chapter.

Assumption 4.1. *The function $f : \mathbb{R}^n \times \mathbb{R}^n \times \mathbb{R}^m \to \mathbb{R}^n$ is continuously differentiable and $f(0, 0, 0) = 0$, i.e., $x_{t,s} = 0$ is an equilibrium of system* (4.1) *for $u_s = 0$.*

Assumption 4.2. *System* (4.1) *has a unique solution for any initial function $\varphi \in \mathcal{C}_\tau$ and any piecewise- and right-continuous input function $u : \mathbb{R}_{\geq 0} \to \mathbb{U}$.*

Assumption 4.3. *The input constraint set $\mathbb{U} \subset \mathbb{R}^m$ is compact and contains the origin in its interior.*

The problem of interest is to stabilize the steady state $x_{t,s} = 0$ via model predictive control.

Remark 4.1. *We do not consider systems with input or measurement delays in this thesis. For these systems, stabilizing MPC schemes can be designed in a straightforward manner by using a forward prediction of the state (Findeisen, 2004, Section 4.5).*

Remark 4.2. *In this chapter, we assume that the full state x_t can be measured. Since the segment $x_t \in \mathcal{C}_\tau$ is infinite-dimensional, this can be restrictive for practical applications.*

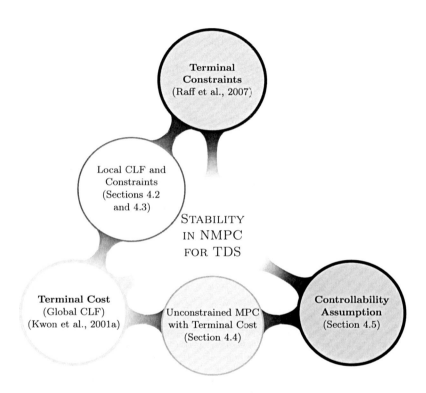

Figure 4.1: Schematic overview of MPC schemes for nonlinear time-delay systems.

4.2 MPC Setup with Terminal Constraints and Asymptotic Stability

The MPC setup considered in this chapter is closely related to the classical schemes for delay-free systems which employ a terminal constraint and a terminal cost, see also Theorem 2.6 and, e.g., the results of Chen and Allgöwer (1998); Mayne et al. (2000). A locally asymptotically stabilizing control law is designed in some neighborhood $\Omega \subseteq \mathcal{C}_\tau$ of the equilibrium. With this locally stabilizing controller, an upper bound on the infinite horizon cost is computed and used as a terminal cost. Furthermore, a constraint is added to the open-loop optimal control problem that requires the final state x_t to lie within the terminal region Ω.

The open-loop finite horizon optimal control problem at sampling instant t_i given the measured state x_{t_i} is formulated as follows.

Problem 4.3.

$$\underset{\bar{u} \in \mathcal{PC}([t_i, t_i+T], \mathbb{R}^m)}{\text{minimize}} \quad J_T(x_{t_i}, \bar{u}) \tag{4.2a}$$

subject to

$$\dot{\bar{x}}(t'; x_{t_i}, t_i) = f(\bar{x}(t'; x_{t_i}, t_i), \bar{x}(t'-\tau; x_{t_i}, t_i), \bar{u}(t')), \qquad t' \in [t_i, t_i + T], \tag{4.2b}$$

$$\bar{x}(t_i + \theta; x_{t_i}, t_i) = x_{t_i}(\theta), \qquad \theta \in [-\tau, 0], \tag{4.2c}$$

$$\bar{u}(t') \in \mathbb{U}, \qquad t' \in [t_i, t_i + T], \tag{4.2d}$$

$$\bar{x}_{t_i+T} \in \Omega, \tag{4.2e}$$

in which

$$J_T(x_{t_i}, \bar{u}) = \int_{t_i}^{t_i+T} F(\bar{x}(t'; x(t_i), t_i), \bar{u}(t')) \, dt' + E(\bar{x}_{t_i+T}).$$

In Problem 4.3, $\bar{x}(t'; x_{t_i}, t_i)$ is the predicted trajectory starting from initial condition x_{t_i} and driven by $\bar{u}(t')$ for $t' \in [t_i, t_i + T]$. The main difference to Problem 2.4 is that the terminal cost E is a functional, and not only a function.

In order to guarantee asymptotic stability in the subsequent analysis, we require the following technical assumption.

Assumption 4.4. *The terminal region $\Omega \subseteq \mathcal{C}_\tau$ is a closed set and contains $0 \in \mathcal{C}_\tau$ in its interior. The terminal cost functional $E : \mathcal{C}_\tau \to \mathbb{R}_{\geq 0}$ is continuously differentiable, positive definite, and there exists a class $\mathcal{K}_\infty$ function $\underline{\alpha}_E : \mathbb{R}_{\geq 0} \to \mathbb{R}_{\geq 0}$ such that $E(x_t) \geq \underline{\alpha}_E(|x(t)|)$. The stage cost $F : \mathbb{R}^n \times \mathbb{U} \to \mathbb{R}_{\geq 0}$ is continuous, $F(0,0) = 0$, and there is a class $\mathcal{K}_\infty$ function $\underline{\alpha}_F : \mathbb{R}_{\geq 0} \to \mathbb{R}_{\geq 0}$ such that*

$$F(x, u) \geq \underline{\alpha}_F(|x|) \quad \text{for all } x \in \mathbb{R}^n, \, u \in \mathbb{U}. \tag{4.3}$$

We assume that the optimal open-loop control which minimizes $J_T(x_{t_i}, \bar{u})$ is given by $u_T^\star(t'; x_{t_i}, t_i)$ for all $t' \in [t_i, t_i + T]$. The associated optimal cost is denoted by $J_T^\star(x_{t_i})$ and the associated predicted trajectory is $x_T^\star(t'; x_{t_i}, t_i)$, $t' \in [t_i, t_i + T]$. For given sampling time δ with $0 < \delta \leq T$, the control input to the system is defined by the following algorithm in the usual receding horizon fashion.

Algorithm 4.4 (Model Predictive Control for Nonlinear Time-Delay Systems). *At each sampling instant* $t_i = i\delta$, $i \in \mathbb{N}_0$, *measure the state* x_{t_i} *and solve Problem 4.3. Apply the input*

$$u_{\mathrm{MPC}}(t) = u_T^*(t; x_{t_i}, t_i), \quad t_i \le t < t_i + \delta. \tag{4.4}$$

to the system until the next sampling instant $t_{i+1} = t_i + \delta$.

The two main assumptions necessary for asymptotic stability of the closed-loop are given in the following.

Assumption 4.5. *The open-loop finite horizon problem 4.3 admits a feasible solution at the initial time* $t = 0$.

Assumption 4.6. *For the nonlinear time-delay system* (4.1), *there exists a locally asymptotically stabilizing controller* $u(t) = k(x_t) \in \mathbb{U}$ *such that the terminal region* Ω *is controlled positively invariant and*

$$\forall x_t \in \Omega : \dot{E}(x_t) \le -F(x(t), k(x_t)). \tag{4.5}$$

We can summarize the main result regarding asymptotic stability of the closed-loop system as follows.

Theorem 4.5 (Stability of MPC for Nonlinear Time-Delay Systems). *Consider the nonlinear time-delay system* (4.1) *and suppose that Assumptions 4.1–4.6 are satisfied. Then, the closed-loop system resulting from the application of the model predictive controller according to Algorithm 4.4 to system* (4.1) *is asymptotically stable. The region of attraction is the set of all initial conditions for which Problem 4.3 is initially feasible.*

Proof. The proof is given in Appendix A.1. □

Note that the previously existing MPC schemes for nonlinear time-delay systems, which either use a global control Lyapunov functional (Kwon et al., 2001a,b; Mahboobi Esfanjani and Nikravesh, 2009a) or an extended zero terminal state constraint (Angrick, 2007; Raff et al., 2007), can be viewed as special cases of this general result. A similar result for instantaneous MPC, i.e., for the limit $\delta \to 0$, has been reported in (Mahboobi Esfanjani, Reble, Münz, Nikravesh, and Allgöwer, 2009). An extension of this result to systems with distributed delay was presented in (Mahboobi Esfanjani and Nikravesh, 2011).

Furthermore note the similarity of Theorem 4.5 for nonlinear time-delay systems to Theorem 2.6 for finite-dimensional continuous-time systems. The main difference is the use of a terminal cost functional instead of a function and the definition of the terminal region Ω. However, the infinite-dimensional nature makes the design of suitable stabilizing design parameters significantly more difficult. Examples for suitable terminal regions and terminal cost functionals will be derived in the following Section 4.3.

4.3 Calculation of the Terminal Region and Terminal Cost

The key element in the stabilizing model predictive control scheme presented in Section 4.2 is a suitable choice of the terminal cost E and the terminal region Ω. The goal of this section is to derive conditions for E and Ω in order to guarantee closed-loop stability under the presented MPC scheme. In this section, we will focus our attention on finding a suitable terminal region Ω and a terminal cost function E such that Assumption 4.6 is satisfied.

Two special cases can be detected for which Assumption 4.6 is directly satisfied. First, the works of Angrick (2007); Raff et al. (2007) consider an extended zero terminal state constraint. Hence, the terminal region only consists of the steady state at the origin, i.e., one single point in the infinite-dimensional space $\mathcal{C}_\tau$. This approach is unattractive from a computational point of view for two reasons. Feasibility problems may occur especially for short prediction horizons because the system has to be steered to the steady state in finite time. In addition, an exact satisfaction of a zero terminal state constraint does require an infinite number of iterations in the numerical optimization. Second, the work of Kwon et al. (2001a,b) uses the whole state space as terminal region and requires the knowledge of a globally stabilizing controller, which might be difficult particularly in the presence of input constraints. In both cases, the invariance of the terminal region is trivially satisfied.

Since it is already a difficult task to calculate a stabilizing control law for linear-time delay systems, we cannot expect to develop a method for general nonlinear time-delay systems, even locally. In this section, we follow the ideas of Chen (1997); Chen and Allgöwer (1998), which have proposed – in the context of MPC for finite-dimensional nonlinear continuous-time systems – to design a linear locally stabilizing control law based on the Jacobi linearization of the nonlinear system. In the second step, a bound on the nonlinearity can be taken into account in order to find a positively invariant region for the nonlinear system, in which Condition (4.5) holds.

The Jacobi linearization of system (4.1) is given by

$$\dot{\tilde{x}}(t) = \tilde{f}(\tilde{x}(t), \tilde{x}(t-\tau), u(t)) = A\tilde{x}(t) + A_\tau \tilde{x}(t-\tau) + Bu(t)\,, \qquad (4.6)$$

in which the matrices are defined by

$$A = \left.\frac{\partial f}{\partial x(t)}\right|_{x_t=0,u=0}\,, \quad A_\tau = \left.\frac{\partial f}{\partial x(t-\tau)}\right|_{x_t=0,u=0}\,, \quad \text{and} \quad B = \left.\frac{\partial f}{\partial u(t)}\right|_{x_t=0,u=0}\,.$$

The difference between the nonlinear system (4.1) and its Jacobi linearization (4.6) will be denoted by Φ in this chapter, i.e.,

$$\Phi(x_t, u(t)) = f(x(t), x(t-\tau), u(t)) - Ax(t) - A_\tau x(t-\tau) - Bu(t)\,. \qquad (4.7)$$

Since f is continuously differentiable and Φ only consists of higher order terms, i.e., it does not contain any linear terms, for any $\gamma \in \mathbb{R}_{>0}$ there exists a $\delta_\gamma \in \mathbb{R}_{>0}$ such that for all $\|x_t\|_\tau \leq \delta_\gamma$ and $|u(t)| < \delta_\gamma$

$$|\Phi(x_t, u(t))| < \gamma\left(|x(t)| + |x(t-\tau)| + |u(t)|\right)\,. \qquad (4.8)$$

In order to design a control law for the Jacobi linearization (4.6), we consider a general linear local control law

$$u(t) = k(x_t) = Kx(t) + \int_{-\tau}^0 K_\tau(\theta)x(t+\theta)d\theta \qquad (4.9)$$

with constant matrix $K \in \mathbb{R}^{m \times n}$ and matrix function $K_\tau(\theta) \in \mathbb{R}^{m \times n}$, together with a quadratic stage cost

$$F(x(t), u(t)) = x(t)^T Q x(t) + u(t)^T R u(t)\,, \qquad (4.10)$$

in which Q and R are symmetric positive definite matrices.

Unfortunately, for nonlinear time-delay systems it is not possible to design the terminal region by exactly following the lines of Chen and Allgöwer (1998) for finite-dimensional systems. The reason for this is as follows. In the delay-free case, it is possible to determine a sufficiently small level set of the positive definite Lyapunov function of the linearized system such that this set is positively invariant also for the nonlinear system. However, for the infinite-dimensional case, even an arbitrarily small level set of a positive definite Lyapunov-Krasovskii functional of the linearized system might not be positively invariant for the nonlinear system. Roughly speaking, this is the case because even for small values of a positive definite functional, the norm of its argument might be arbitrarily large. This is similar to the well-known fact that there is no equivalence between different norms in infinite-dimensional spaces. We illustrate this in the following brief example.

Example 4.6. *Consider the simple scalar nonlinear time-delay system*

$$\dot{x}(t) = x(t-\tau)^4 + u(t) \tag{4.11}$$

with constant time-delay $\tau = 1$. The Jacobi linearization of system (4.11) about the origin is $\dot{x}(t) = u(t)$ and the closed-loop resulting from the application of the simple linear control law $u(t) = k(x_t) = -2x(t)$ to the Jacobi linearization is asymptotically stable. For instance, this can be shown by using Theorem 2.9 and the following Lyapunov-Krasovskii functional and its derivative along trajectories of the linearized closed-loop

$$E(x_t) = x(t)^2 + \int\limits_{-\tau}^{0} x(t+\theta)^2 \, d\theta \,, \tag{4.12a}$$

$$\dot{E}(x_t) = 2x(t)u(t) + x(t)^2 - x(t-\tau)^2 = -3x(t)^2 - x(t-\tau)^2 \,. \tag{4.12b}$$

Since the Jacobi linearization is asymptotically stable, we know that the closed-loop consisting of the original nonlinear system (4.11) and the linear control law is also locally asymptotically stable (Kolmanovskii and Myshkis, 1999). Following the lines of the scheme presented in Chen and Allgöwer (1998) for delay-free systems, a natural choice for the terminal region would be

$$\Omega = \{x_t \in \mathcal{C}_\tau \,:\, E(x_t) \leq \alpha\} \,, \qquad \alpha \in \mathbb{R}_{>0} \,.$$

In the delay-free case, choosing $\alpha \in \mathbb{R}_{>0}$ sufficiently small allows to guarantee positive invariance of Ω and satisfaction of (4.5) for the original nonlinear system (Chen and Allgöwer, 1998). However, this is not possible in the current example. For a given $\alpha \in \mathbb{R}_{>0}$, define $\xi = \min\{\sqrt{\alpha}/2, 2/3\}$ and $\mathfrak{t} = \min\{\alpha\xi/2, \tau/2\}$ and consider $\widehat{x}_t \in \Omega$ given by

$$\widehat{x}_t(\theta) = \max\left\{\frac{-1}{\xi\mathfrak{t}}(\theta+\tau-\mathfrak{t}), \xi\right\} = \begin{cases} \frac{-1}{\xi\mathfrak{t}}(\theta+\tau-\mathfrak{t}), & \theta \leq -\tau+\mathfrak{t} \\ \xi, & \theta > -\tau+\mathfrak{t} \end{cases} \,.$$

See Figure 4.2 for a sketch of $\widehat{x}_t$ and note that direct calculations reveal that indeed $E(\widehat{x}_t) \leq \alpha$. By taking $\xi \leq 2/3$ into account, we obtain for the derivative of (4.12a) along trajectories of the nonlinear closed-loop

$$\dot{E}(\widehat{x}_t) = 2\widehat{x}(t)\left(\widehat{x}(t-\tau)^4 + u(t)\right) + \widehat{x}(t)^2 - \widehat{x}(t-\tau)^2 = -3\xi^2 - \frac{1}{\xi^2}\left(1 - \frac{2}{\xi}\right) > 0 \,.$$

Hence, Condition (4.5) in Assumption 4.6 cannot be guaranteed by choosing α small enough. Similarly, positive invariance of Ω is not necessarily given.

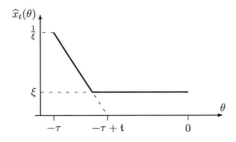

Figure 4.2: Sketch of $\widehat{x}_t \in \Omega$ used in Example 4.6.

In the following sections, we present different design schemes in order to obtain a suitable terminal cost functional E and terminal region Ω using the Jacobi linearization about the origin. The first scheme presented in Section 4.3.1 only requires a locally stabilizing linear control law, but results in a complicated terminal region. The second scheme uses an additional Razumikhin condition, which allows the calculation of a more simple terminal region as shown in Section 4.3.2. When considering a terminal cost functional motivated by the use of the Razumikhin condition and an additional condition on the sampling time, the terminal region can be defined as a sublevel set of this terminal cost as derived in Section 4.3.3. In Section 4.3.4, we show that the condition on the sampling time can be removed by including an additional exponential weighting term in the terminal cost. The four different schemes are compared in Section 4.3.5.

4.3.1 General Linearization-based Design

In this section, we consider the least restrictive possible design scheme based on the Jacobi linearization. Similar to the results for delay-free systems presented in Chen and Allgöwer (1998), it is shown that each nonlinear time-delay system, which possesses a stabilizable Jacobi linearization about the origin, can be stabilized by MPC with a quadratic terminal cost functional and a finite terminal region. In contrast to the delay-free case, the terminal region is not defined as a sublevel set of the terminal cost, but instead as the intersection of such a sublevel set with a sphere in the infinite-dimensional space $\mathcal{C}_\tau$ defined by the norm $\| \cdot \|_\tau$.

Using the definitions given so far, we can summarize the main result of this section as follows.

Theorem 4.7 (Design for Time-Delay Systems with Stabilizable Linearization). *Consider the nonlinear time-delay system* (4.1) *and the quadratic stage cost* (4.10). *Suppose that Assumptions 4.1–4.3 are satisfied and that there exists a linear local control law* (4.9) *such that the linearized system* (4.6) *is asymptotically stable. Then, there exists a terminal cost* *functional*

$$E(x_t) = x(t)^T P_0 x(t) + \int_{-\tau}^{0} \int_{-\tau}^{0} x(t+\theta_1)^T P_3(\theta_1, \theta_2) x(t+\theta_2) d\theta_1 d\theta_2$$

$$+ 2 x(t)^T \int_{-\tau}^{0} P_1(\theta) x(t+\theta) d\theta + \int_{-\tau}^{0} x(t+\theta)^T P_2(\theta) x(t+\theta) d\theta \qquad (4.13)$$

with symmetric positive definite matrix $P_0 = P_0^T \in \mathbb{R}^{n \times n}$ *and matrix functions* $P_1(\theta) \in \mathbb{R}^{n \times n}$, $P_2(\theta) = P_2(\theta)^T \in \mathbb{R}^{n \times n}$, $P_3(\theta_1, \theta_2) = P_3(\theta_2, \theta_1)^T \in \mathbb{R}^{n \times n}$, *and there exists a terminal region*

$$\Omega = \left\{ x_t \in \mathcal{C}_\tau \ : \ E(x_t) \leq \mu \frac{\alpha^2}{4}, \quad \|x_t\|_\tau \leq \frac{\alpha}{2} \right\} \qquad (4.14)$$

in which $\alpha \in \mathbb{R}_{>0}$ *and* $\mu \in \mathbb{R}_{>0}$, *such that Assumptions 4.4 and 4.6 are satisfied. Furthermore, the closed-loop resulting from the application of the model predictive controller according to Algorithm 4.4 to system* (4.1) *is asymptotically stable.*

Proof. Since the closed-loop consisting of the Jacobi linearization (4.6) in combination with the linear control law $u(t) = k(x_t)$, i.e.,

$$\dot{\tilde{x}}(t) = \tilde{A}\tilde{x}(t) + A_\tau \tilde{x}(t-\tau) + Bk(\tilde{x}_t), \qquad (4.15)$$

is asymptotically stable, there exists a quadratic Lyapunov-Krasovskii functional $\tilde{E}(x_t)$ of the form (4.13) whose derivative along trajectories of (4.15) satisfies

$$\dot{\tilde{E}}(x_t) \leq -\varepsilon_1 |x(t)|^2 - \varepsilon_2 |x(t-\tau)|^2 - \varepsilon_3 \int_{-\tau}^{0} |x(t+\theta)|^2 d\theta$$

with constants $\varepsilon_1, \varepsilon_2, \varepsilon_3 \in \mathbb{R}_{>0}$, see Kharitonov and Zhabko (2003). The derivative of $\tilde{E}$ along trajectories of the nonlinear system (4.1) satisfies

$$\dot{\tilde{E}}(x_t) \leq -\varepsilon_1 |x(t)|^2 - \varepsilon_2 |x(t-\tau)|^2 - \varepsilon_3 \int_{-\tau}^{0} |x(t+\theta)|^2 d\theta$$

$$+ 2\Phi(x_t, u(t))^T \left(P_0 x(t) + \int_{-\tau}^{0} P_1(\theta) x(t+\theta) d\theta \right).$$

Due to Inequality (4.8) and since $2ab \leq a^2 + b^2$ for all $a, b \in \mathbb{R}$,

$$2\Phi(x_t, u(t))^T P_0 x(t) \leq \gamma \|P_0\| \left(4|x(t)|^2 + |x(t-\tau)|^2 + |u(t)|^2 \right)$$

holds for all $\|x_t\|_\tau \leq \delta_\gamma$ and $|u(t)| \leq \delta_\gamma$. Similarly, using the Cauchy-Schwarz inequality (Bronstein et al., 2000) the following inequalities hold

$$2\Phi(x_t, u(t))^T \int_{-\tau}^{0} P_1(\theta) x(t+\theta) d\theta$$

$$\leq \gamma \left(|x(t)|^2 + |x(t-\tau)|^2 + |u(t)|^2 \right) + 3\gamma \left| \int_{-\tau}^{0} P_1(\theta) x(t+\theta) d\theta \right|^2$$

$$\leq \gamma \left(|x(t)|^2 + |x(t-\tau)|^2 + |u(t)|^2 \right) + 3\gamma \tau \|P_1\|_\tau^2 \int_{-\tau}^{0} |x(t+\theta)|^2 d\theta$$

for $\|x_t\|_\tau < \delta_\gamma$ and $|u(t)| < \delta_\gamma$ with $\|P_1\|_\tau = \sup_{\theta \in [-\tau, 0]} \|P_1(\theta)\|$. Combining these findings yields

$$\dot{\tilde{E}}(x_t) \leq -(\varepsilon_1 - 4\gamma\|P_0\| - \gamma)\,|x(t)|^2 - (\varepsilon_2 - \gamma\,\|P_0\| - \gamma)\,|x(t-\tau)|^2$$
$$- (\varepsilon_3 - 3\gamma\tau\,\|P_1\|_\tau^2)\int_{-\tau}^0 |x(t+\theta)|^2 d\theta + \gamma(\|P_0\| + 1)\,|u(t)|^2\,.$$

Furthermore, we obtain for the linear control law $u(t) = k(x_t)$

$$|u(t)|^2 = |k(x_t)|^2 \leq 2\|K^T K\|\,|x(t)|^2 + 2\tau\,\|K_\tau\|_\tau^2 \int_{-\tau}^0 |x(t+\theta)|^2 d\theta \qquad (4.16)$$

with $\|K_\tau\|_\tau = \sup_{\theta \in [-\tau, 0]} \|K_\tau(\theta)\|$ and, hence,

$$\dot{\tilde{E}}(x_t) \leq -(\varepsilon_1 - 4\gamma\|P_0\| - \gamma - 2\gamma(\|P_0\| + 1)\|K^T K\|)\,|x(t)|^2$$
$$- (\varepsilon_2 - \gamma\,\|P_0\| - \gamma)\,|x(t-\tau)|^2$$
$$- (\varepsilon_3 - 3\gamma\tau\,\|P_1\|_\tau^2 - 2\tau\,\gamma(\|P_0\| + 1)\|K_\tau\|_\tau^2)\int_{-\tau}^0 |x(t+\theta)|^2 d\theta$$

for $|k(x_t)|^2 < \delta_\gamma$. It is clearly possible to choose $\gamma, \beta \in \mathbb{R}_{>0}$ such that the following three inequalities are satisfied

$$\beta\,(\varepsilon_1 - 4\gamma\|P_0\| - \gamma - 2\gamma(\|P_0\| + 1)\|K^T K\|) > \|Q + 2K^T RK\|\,,$$
$$(\varepsilon_2 - \gamma\,\|P_0\| - \gamma) > 0\,,$$
$$\beta\,(\varepsilon_3 - 3\gamma\tau\,\|P_1\|_\tau^2 - 2\tau\,\gamma(\|P_0\| + 1)\|K_\tau\|_\tau^2) > 2\tau\|R\|\,\|K_\tau\|_\tau^2\,.$$

Now define the terminal region Ω as in (4.14) with $\alpha > 0$ chosen such that $\alpha \leq 2\,\delta_\gamma$ and such that $x_t \in \Omega$ implies $k(x_t) \in \mathcal{U}$ and $|k(x_t)|^2 < \delta_\gamma$. The satisfaction of both conditions is always possible for some small enough $\alpha > 0$. Moreover, define the terminal cost functional $E(x_t) = \beta\,\tilde{E}(x_t)$. Then, clearly all x_t with $\|x_t\|_\tau \leq \alpha$ and, consequently, all $x_t \in \Omega$ satisfy

$$\dot{E}(x_t) \leq -x(t)^T Q x(t) - u(t)^T R u(t) \qquad (4.17)$$

when using the local control law $u(t) = k(x_t)$. Furthermore, the terminal region Ω is positively invariant for the choice of $\mu = \beta\,\lambda_{\min}(P_0)$. This can be shown, similar to the proofs in (Melchor-Aguilar and Niculescu, 2007; Reble and Allgöwer, 2010a), by contradiction using $\forall x_t \in \Omega\,:\,\dot{E}(x_t) \leq 0$ and by noting that $E(x_t) \geq \beta\,\lambda_{\min}(P_0)|x(t)|^2$. Without loss of generality assume that $x_{t_0} \in \Omega$. For the sake of contradiction, assume that Ω is not positively invariant. Since $x(t)$ is a continuous function of time, there exists a $t_1 > t_0$ for which $x_{t_1} \notin \Omega$ and $\|x_t\|_\tau < \frac{3\alpha}{4}$ for all $t \leq t_1$. Note that $\dot{E}(x_t) < 0$ for all x_t with $\|x_t\|_\tau < \frac{3\alpha}{4}$ as shown in the first part of this proof. Thus, $E(x_{t_1}) \leq E(x_{t_0})$. However, this implies $\|x_{t_1}\|_\tau > \frac{\alpha}{2}$ because we assume $x_{t_1} \notin \Omega$. It follows that there is a time t_2 with $t_0 < t_2 \leq t_1$ for which $|x(t_2)| > \frac{\alpha}{2}$, and $E(x_{t_2}) \leq E(x_{t_0})$ because of $\dot{E} < 0$. Using the aforementioned lower bound on E, we directly obtain

$$E(x_{t_2}) \geq \beta\,\lambda_{\min}(P_0)|x(t_2)|^2 > \beta\,\lambda_{\min}(P_0)\frac{\alpha^2}{4}\,.$$

But this implies $E(x_{t_0}) > \mu \frac{\alpha^2}{4}$, which contradicts the assumption $x_{t_0} \in \Omega$. Hence, the terminal region Ω is positively invariant.

Combining (4.17) and the positive invariance shows that Assumption 4.6 is satisfied. The satisfaction of Assumption 4.4 is straightforward due to the definitions of F, E, and Ω, respectively. Asymptotic stability follows directly by use of Theorem 4.5, for which all necessary assumptions are satisfied. $\qquad\square$

Remark 4.8. *The standard converse Lyapunov Theorem for linear time-delay systems (Gu et al., 2003, Proposition 7.4) only guarantees the existence of a complete quadratic functional $E(x_t)$ with $\dot{E}(x_t) \leq -\varepsilon_1 |x(t)|^2$, which is not sufficient to ensure $\dot{E}(x_t) \leq -F(x(t), k(x_t))$ for a general linear control law of the form (4.9). Thus, the more general result of Kharitonov and Zhabko (2003) is needed in the proof of Theorem 4.7.*

The result is based on rather mild assumptions since only a stabilizable Jacobi linearization is required. However, the resulting terminal region as defined in (4.14) is quite complicated.

Example 4.9. *We consider again Example 4.6. We choose the stage cost $F(x,u) = x^2 + 0.5u^2$ and use the terminal cost functional $E(x_t)$ as defined in (4.12a) together with the locally stabilizing linear control law $u(t) = k(x_t) = -2x(t)$. Straightforward manipulations show that the derivative along trajectories of the nonlinear closed-loop satisfies*

$$\dot{E}(x_t) = -3x(t)^2 - x(t-\tau)^2 + 2x(t)x(t-\tau)^4 \leq -3x(t)^2 = -F(x(t), k(x_t))$$

for all $x_t \in \mathcal{C}_\tau$ which satisfy $\|x_t\|_\tau \leq \frac{\alpha}{2}$ with $\alpha = 1$. For $\mu = 1$, the terminal region Ω defined in (4.14) is positively invariant, which can be shown along the lines of the proof of Theorem 4.7. Hence, MPC according to Algorithm 4.4 with design parameters E and Ω asymptotically stabilizes the origin of the closed-loop.

Design using LMIs

In this section, we provide exemplary conditions for the local linear control law $u(t) = k(x_t)$ in terms of linear matrix inequalities (LMIs).

Theorem 4.10 (LMI Condition for Local Control Law). *Consider the nonlinear time-delay system (4.1) and the quadratic stage cost (4.10). Suppose that Assumptions 4.1–4.3 are satisfied. If there exist symmetric matrices $\Lambda \succ 0$, $\Upsilon \succ 0$, a matrix Γ and a constant positive scalar $\varepsilon \in \mathbb{R}_{>0}$ solving the following LMI*

$$\begin{bmatrix} \Xi_2 + \Upsilon + \varepsilon I & A_\tau \Lambda & \vdots & \Lambda Q^{1/2} & \Gamma^T R^{1/2} \\ \star & -\Upsilon + \varepsilon I & \vdots & 0 & 0 \\ \hdashline \star & \star & \vdots & -I & 0 \\ \star & \star & \vdots & \star & -I \end{bmatrix} \prec 0 \qquad (4.18)$$

in which $\Xi_2 = \Lambda A^T + A\Lambda + \Gamma^T B^T + B\Gamma$, then the control law $u(t) = Kx(t)$ with $K = \Gamma \Lambda^{-1}$ locally asymptotically stabilizes the nonlinear time-delay system (4.1). Furthermore, consider the cost functional E given by

$$E(x_t) = x(t)^T P x(t) + \int_{-\tau}^{0} x^T(t+\theta) S x(t+\theta) d\theta \qquad (4.19)$$

with parameters $P = \Lambda^{-1}$ and $S = \Lambda^{-1}\Upsilon\Lambda^{-1}$ and the terminal region defined by

$$\Omega = \left\{ x_t \ : \ E(x_t) \leq \frac{\lambda_{\min}(P)\,\delta_\gamma^2}{4}, \quad \|x_t\|_\tau \leq \frac{\delta_\gamma}{2} \right\}, \tag{4.20}$$

with $\gamma \in \mathbb{R}_{>0}$ chosen small enough such that

$$\gamma \leq \varepsilon \, \frac{\lambda_{\min}(P)^2}{2\,\lambda_{\max}(P)} \tag{4.21}$$

and small enough such that $|x| < \frac{\delta_\gamma}{2} \Rightarrow u = Kx \in \mathcal{U}$. Then, Assumption 4.6 is satisfied.

Proof. The proof is given in Appendix A.2. □

4.3.2 Combination of Lyapunov-Krasovskii and Lyapunov-Razumikhin

In this section, we use additional assumptions on the local stabilizing control law in order to obtain a simpler terminal region. To this end, we do not only require the existence of a locally stabilizing linear control law as in Section 4.3.1, but assume this local control law to satisfy a Lyapunov-Razumikhin condition as stated in the following assumption.

Assumption 4.7 (Razumikhin Condition). *There exists a linear local control law $u(t) = k(x_t) = Kx(t) + \int_{-\tau}^{0} K_\tau(\theta)x(t+\theta)d\theta$ and constants $\varepsilon \in \mathbb{R}_{>0}$, $\rho \in \mathbb{R}_{>1}$ such that the derivative of the Lyapunov-Razumikhin function $V(x(t)) = x(t)^T P x(t)$ along trajectories of the linearized system (4.6) satisfies $\dot{V}(x(t)) \leq -\varepsilon\,|x(t)|^2$ whenever*

$$\forall \theta \in [-\tau, 0] \ : \ V(x(t+\theta)) \leq \rho\,V(x(t)). \tag{4.22}$$

With this assumption, we can design a terminal region and terminal cost as summarized in the following theorem.

Theorem 4.11 (Design using Lyapunov-Krasovskii and Lyapunov-Razumikhin Arguments). *Consider the nonlinear time-delay system (4.1) and the quadratic stage cost (4.10). Suppose that Assumptions 4.1–4.3 are satisfied and that Assumption 4.7 holds. Then, there exist a terminal cost functional of the form (4.13) and a terminal region*

$$\Omega = \left\{ x_t \in \mathcal{C}_\tau \ : \ \max_{\theta \in [-\tau, 0]} V(x(t+\theta)) \leq \alpha \right\}, \tag{4.23}$$

in which $\alpha \in \mathbb{R}_{>0}$, such that Assumptions 4.4 and 4.6 are satisfied. Furthermore, the closed-loop resulting from the application of the model predictive controller according to Algorithm 4.4 to system (4.1) is asymptotically stable.

Proof. In the first part of the proof, it is shown that there is a sufficiently small $\alpha \in \mathbb{R}_{>0}$ such that for all $x_t \in \Omega$ the derivative of the Lyapunov-Razumikhin function V along trajectories of the nonlinear system (4.1) satisfies

$$\dot{V}(x(t)) \leq -\frac{\varepsilon}{2}\,|x(t)|^2 \quad \text{whenever } \forall \theta \in [-\tau, 0] \ : \ V(x(t+\theta)) \leq \rho\,V(x(t)). \tag{4.24}$$

To this end, note that if $\|x_t\|_\tau \leq \delta_\gamma$, $|u(t)| < \delta_\gamma$, and $\forall \vartheta \in [-\tau, 0] : V(x(t+\vartheta)) \leq \rho V(x(t))$, then

$$
\begin{aligned}
\dot{V}(x(t)) &\leq -\varepsilon |x(t)|^2 + 2x(t)^T P \Phi(x_t, u(t)) \\
&\leq -\varepsilon |x(t)|^2 + 2\gamma |x(t)| \, \|P\| \, (|x(t)| + |x(t-\tau)| + |u(t)|) \\
&\leq -(\varepsilon - 4\gamma \|P\|) \, |x(t)|^2 + \gamma \|P\| \, |x(t-\tau)|^2 + \gamma \|P\| \, |u(t)|^2 \\
&\overset{(4.16)}{\leq} -(\varepsilon - 4\gamma \|P\| - 2\gamma \|P\| \, \|K^T K\|) \, |x(t)|^2 \\
&\qquad + \gamma \|P\| \, |x(t-\tau)|^2 + 2\tau \gamma \|P\| \, \|K_\tau\|_\tau^2 \int_{-\tau}^0 |x(t+\vartheta)|^2 d\vartheta \\
&\leq \left(-\varepsilon + \gamma \|P\| \left(4 + 2 \|K^T K\| + (1 + 2\tau^2 \|K_\tau\|_\tau^2) \, \rho \frac{\lambda_{\max}(P)}{\lambda_{\min}(P)} \right) \right) |x(t)|^2 .
\end{aligned}
$$

In the last inequality, we explicitly used that $V(x(t+\vartheta)) \leq \rho V(x(t))$ for all $\vartheta \in [-\tau, 0]$. Now choose $\gamma \in \mathbb{R}_{>0}$ such that

$$
\gamma \|P\| \left(4 + 2 \|K^T K\| + (1 + 2\tau^2 \|K_\tau\|_\tau^2) \, \rho \frac{\lambda_{\max}(P)}{\lambda_{\min}(P)} \right) \leq \frac{\varepsilon}{2} ,
$$

and $\alpha \in \mathbb{R}_{>0}$ such that (i) $\alpha < \lambda_{\min}(P) \, \delta_\gamma$, and (ii) $x_t \in \Omega$ implies $k(x_t) \in \mathcal{U}$ and $|k(x_t)|^2 < \delta_\gamma$, which is always possible for α small enough. Then, the condition $\dot{V}(x(t)) \leq -\varepsilon/2 \, |x(t)|^2$ holds for all $x_t \in \Omega$ whenever $V(x(t+\vartheta)) \leq \rho V(x(t))$ for all $\vartheta \in [-\tau, 0]$. Furthermore, the terminal region Ω is controlled positively invariant when applying the local control law $k(x_t)$.

Assumption 4.7 directly implies that V is a Lyapunov-Razumikhin function for the linearized system with control law $u(t) = k(x_t)$. Hence, the linearized system is asymptotically stable when using this local control law. Consequently, as shown in the proof of Theorem 4.7, there exist a quadratic cost functional of the form (4.13) and an $\alpha \in \mathbb{R}_{>0}$ small enough such that $\dot{E}(x_t) \leq -x(t)^T Q x(t) - u(t)^T R u(t)$ for all $x_t \in \Omega$ by using the local control law $u(t) = k(x_t)$, see (4.17).

Therefore, the local control law $k(x_t)$ satisfies Condition (4.5) in Assumption 4.6. Furthermore, the terminal region Ω is positively invariant when applying $k(x_t)$ as shown in the first part of the proof. Assumption 4.4 is directly satisfied due to the definitions of F, E, and Ω, respectively. With this, all assumptions necessary for Theorem 4.5 are satisfied. Hence, asymptotic stability is guaranteed. $\qquad \square$

The Razumikhin condition in Assumption 4.7 is more restrictive than the existence of a locally stabilizing linear control law as required in Theorem 4.7. However, the terminal region (4.23) is of simpler form than the terminal region for the general design (4.14).

Example 4.12. *We consider again Example 4.6 with the stage cost $F(x, u) = x^2 + 0.5u^2$. It is simple to show that Assumption 4.7 is satisfied for the locally stabilizing control law $u(t) = k(x_t) = -2x(t)$ and the Lyapunov-Razumikhin function $V(x) = x^2$. The derivative of V along trajectories of the closed-loop consisting of the local control law $u(t) = k(x_t) = -2x(t)$ and the nonlinear time-delay system (4.11) satisfies the Razumikhin-type condition*

$$
\dot{V}(x(t)) \leq -3 \, |x(t)|^2 \qquad \text{whenever} \quad \forall \theta \in [-\tau, 0] : V(x(t+\theta)) \leq 2 V(x(t)) \qquad (4.25)
$$

for all x_t which satisfy $\|x_t\|_\tau \leq \frac{1}{2}$. Furthermore, the terminal cost functional $E(x_t)$ given in (4.12a) satisfies $\dot{E}(x_t) \leq -F(x(t), k(x_t))$ for these x_t as shown in Example 4.9. Hence, the terminal region $\Omega = \left\{ x_t \in \mathcal{C}_\tau \; : \; \max_{\theta \in [-\tau,0]} V(x(t+\theta)) \leq \frac{1}{4} \right\}$ and the terminal cost E satisfy Assumption 4.6. Hence, MPC according to Algorithm 4.4 with design parameters E and Ω asymptotically stabilizes the origin.

Design using LMIs

In this section, we provide exemplary conditions in terms of LMIs for the local linear control law $u(t) = k(x_t)$ used in Theorem 4.11. In contrast to Section 4.3.1, the LMIs used to ensure satisfaction of Assumption 4.6 can be separated into two parts: the first set of LMIs guarantees invariance of the terminal region, see Lemma 4.13, and the second set ensures the Lyapunov condition (4.5), see Lemma 4.14.

The following lemma provides LMI conditions for the controlled invariance of a certain terminal region Ω.

Lemma 4.13 (LMI Condition for Invariance of Ω). *Consider the nonlinear time-delay system (4.1) and the quadratic stage cost (4.10). Suppose that Assumptions 4.1–4.3 are satisfied. If there exist symmetric matrices $\Lambda \succ 0$, $\Lambda_i \succ 0$, $i = 1, 2, 3$ and a matrix Γ of appropriate dimensions solving the following LMIs*

$$
\begin{bmatrix}
\Xi_1 + 2\tau\Lambda & \tau A_\tau(A\Lambda + B\Gamma) & \tau A_\tau^2 \Lambda & \tau A_\tau \Lambda \\
\star & -\tau\Lambda_1 & 0 & 0 \\
\star & \star & -\tau\Lambda_2 & 0 \\
\star & \star & \star & -\tau\Lambda_3
\end{bmatrix} \prec 0 \tag{4.26a}
$$

$$
\Lambda_i - \Lambda \prec 0, \quad i = 1, 2, \tag{4.26b}
$$

in which $\Xi_1 = \Lambda(A + A_\tau)^T + (A + A_\tau)\Lambda + \Gamma^T B^T + B\Gamma$, then there exists $\alpha \in \mathbb{R}_{>0}$ such that the local control law $u(t) = Kx(t)$ with $K = \Gamma\Lambda^{-1}$ renders the terminal region

$$
\Omega = \left\{ x_t \; : \; \max_{\theta \in [-\tau,0]} x(t+\theta)^T P x(t+\theta) \leq \alpha \right\} \tag{4.27}
$$

positively invariant for $P = \Lambda^{-1}$.

Proof. The proof is given in Appendix A.3 and contains an implicit formula for α given by (A.22). □

In the next lemma, we derive LMI conditions for the terminal cost functional E.

Lemma 4.14 (LMI Condition for Terminal Cost Functional). *Consider the nonlinear time-delay system (4.1) and the quadratic stage cost (4.10). Suppose that Assumptions 4.1–4.3 are satisfied. If there exist symmetric matrices $\Lambda \succ 0$, $\Upsilon \succ 0$, a matrix Γ and a constant positive scalar $\varepsilon \in \mathbb{R}_{>0}$ solving the following LMI*

$$
\left[
\begin{array}{cc|cc}
\Xi_2 + \Upsilon + \varepsilon I & A_\tau\Lambda & \Lambda Q^{1/2} & \Gamma^T R^{1/2} \\
\star & -\Upsilon + \varepsilon I & 0 & 0 \\
\hline
\star & \star & -I & 0 \\
\star & \star & \star & -I
\end{array}
\right] \prec 0 \tag{4.28}
$$

in which $\Xi_2 = \Lambda A^T + A\Lambda + \Gamma^T B^T + B\Gamma$, then there exists $\alpha \in \mathbb{R}_{>0}$ such that the local control law $u(t) = Kx(t)$ with $K = \Gamma \Lambda^{-1}$ ensures $\dot{E}(x_t) \leq -F(x(t), Kx(t))$ for all $x_t \in \Omega$. Herein, Ω is defined in (4.27) and the terminal cost functional E is defined in (4.19) with parameters $P = \Lambda^{-1}$ and $S = \Lambda^{-1}\Upsilon\Lambda^{-1}$.

Proof. The proof is given in Appendix A.4 and contains an implicit condition on α given by (A.23). □

Combining the previous results from this section, we directly obtain the following theorem.

Theorem 4.15 (LMI Condition for Local Control Law). *Consider the nonlinear time-delay system (4.1) and the quadratic stage cost (4.10). Suppose that Assumptions 4.1–4.3 are satisfied. If there exist symmetric positive definite matrices $\Lambda, \Lambda_1, \Lambda_2, \Lambda_3, \Upsilon$, and a matrix Γ such that LMIs (4.26) and (4.28) admit a feasible solution, then there exists $\alpha \in \mathbb{R}_{>0}$ small enough such that terminal region Ω and terminal cost functional E defined in (4.27) and (4.19), respectively, with parameters $P = \Lambda^{-1}$ and $S = \Lambda^{-1}\Upsilon\Lambda^{-1}$, satisfy Assumption 4.6.*

Proof. The input constraint set $\mathbb{U}$ contains the origin in its interior. Therefore, the proof of the theorem directly follows from Lemmata 4.13 and 4.14 because it is always possible to choose α small enough such that the input constraints are satisfied by the local control law $u(t) = Kx(t)$ for all $x_t \in \Omega$. □

LMI (4.18) in Theorem 4.10 is less conservative than the LMI condition in Theorem 4.15 in Section 4.3.2 in the sense that if there is a solution to the conditions in Theorem 4.15, then the assumptions in Theorem 4.10 are satisfied. This can be easily seen because LMI (4.18) directly relates to the Lyapunov-Krasovskii condition (4.28) for the terminal cost functional in Lemma 4.14 and an additional LMI (4.26) is required in Theorem 4.15. However, the terminal region (4.20) is more complicated and might make the numerical solution of the open-loop optimal control problem more difficult.

4.3.3 Design by Lyapunov-Razumikhin Arguments

The terminal regions defined in the preceding Sections 4.3.1 and 4.3.2 are not defined as sublevel sets of the quadratic terminal cost functional, which is in contrast to most literature concerning the delay-free case, in particular all schemes based on the early results of Chen and Allgöwer (1998). Indeed, such a sublevel set is not necessarily positively invariant along trajectories of the closed-loop. However, several results from nonlinear delay-free systems heavily rely on the definition of the terminal region as a sublevel set of the terminal cost, such as unconstrained MPC schemes (Graichen and Kugi, 2010; Hu and Linnemann, 2002; Limon et al., 2006). For this reason, an alternative scheme is presented in this section which allows such a definition under additional assumptions.

In Section 4.3.2 a terminal region of the form (4.23) has been used. Using this region with the respective parameter α chosen properly, positive invariance and satisfaction of Condition (4.5) in Assumption 4.6 can be shown for the nonlinear time-delay system. An obvious choice for the terminal cost functional is to consider $E(x_t) = \max\limits_{\theta \in [-\tau, 0]} V(x(t+\theta))$.

However, it is clear that $E(x_t)$ is constant on certain intervals and a decrease of E can only be guaranteed after the time-delay τ. See Figure 4.3 for an illustrating sketch. Hence, (4.5)

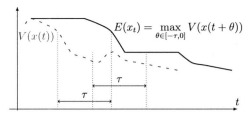

Figure 4.3: Sketch on the necessity of Assumption 4.9 for the design based on Lyapunov-Razumikhin arguments in Section 4.3.3. In general, the terminal cost functional $E(x_t) = \max_{\theta \in [-\tau, 0]} V(x(t + \theta))$ is constant on certain intervals.

cannot be satisfied. For this reason, we replace Assumption 4.6 by Assumption 4.8. We can show that this slightly less restrictive assumption is still sufficient for the assertion of Theorem 4.5 as stated in Theorem 4.16.

Assumption 4.8. *Assumption 4.6 is satisfied with* (4.5) *replaced by the less restrictive condition*

$$\forall x_t \in \Omega \; : \; E(x_{t+\delta}) - E(x_t) \leq - \int_t^{t+\delta} F(x(t'), k(x_{t'})) \, dt' . \tag{4.29}$$

Note that Assumption 4.6 implies Assumption 4.8, but not vice versa. Theorem 4.5 can be generalized as follows.

Theorem 4.16 (Stability of MPC for Nonlinear Time-Delay Systems). *Consider the nonlinear time-delay system* (4.1) *and suppose that Assumptions 4.1–4.5 and 4.8 are satisfied. Then, the closed-loop system resulting from the application of the model predictive controller according to Algorithm 4.4 to system* (4.1) *is asymptotically stable. The region of attraction is the set of all initial conditions for which Problem 4.3 is initially feasible.*

Proof. The proof is exactly the same as the proof of Theorem 4.5 given in Section A.1. Careful inspection of the proof reveals that (4.5) is only used to show (4.29), which is used in the remainder of the proof. □

A decrease of E can only be guaranteed after the time-delay τ, which motivates to use the following additional assumption.

Assumption 4.9. *The sampling time of the MPC algorithm 4.4 is strictly larger than the time-delay, i.e., $\delta > \tau$.*

In the following, the goal is to find conditions on the terminal cost functional E and the terminal region Ω, defined as a sublevel set of E, such that Ω is positively invariant and Condition (4.29) is satisfied. To this end, we consider a local linear control law satisfying the Razumikhin-type condition given in Assumption 4.7, which has also been used in Section 4.3.2.

Theorem 4.17 (Design using Lyapunov-Razumikhin Arguments). *Consider the nonlinear time-delay system (4.1) and the quadratic stage cost (4.10). Suppose that Assumptions 4.1– 4.3 are satisfied and that Assumptions 4.7 and 4.9 hold. Then, there exist a terminal cost functional of the form*

$$E(x_t) = \beta \max_{\theta \in [-\tau, 0]} V(x(t + \theta)) \tag{4.30}$$

and a terminal region

$$\Omega = \left\{ x_t \ : \ \max_{\theta \in [-\tau, 0]} V(x(t + \theta)) \leq \alpha \right\}, \tag{4.31}$$

in which $\alpha, \beta \in \mathbb{R}_{>0}$, such that Assumptions 4.4 and 4.8 are satisfied. Furthermore, the closed-loop resulting from the application of the model predictive controller according to Algorithm 4.4 to system (4.1) is asymptotically stable.

Proof. In the first part of the proof, it is shown that there is a sufficiently small $\alpha \in \mathbb{R}_{>0}$ such that for all $x_t \in \Omega$ the derivative of V along trajectories of the nonlinear system (4.1) satisfies $\dot{V}(x(t)) \leq -\varepsilon/2 \, |x(t)|^2$ whenever Condition (4.22) holds. To this end, we can follow the first part of the proof of Theorem 4.11 to show

$$\dot{V}(x(t)) \leq \left(-\varepsilon + \gamma \|P\| \left(4 + 2 \|K^T K\| + (1 + 2\tau^2 \|K_\tau\|_\tau^2) \rho \frac{\lambda_{\max}(P)}{\lambda_{\min}(P)} \right) \right) |x(t)|^2.$$

It is now possible to choose $\gamma \in \mathbb{R}_{>0}$ such that

$$\gamma \|P\| \left(4 + 2 \|K^T K\| + (1 + 2\tau^2 \|K_\tau\|_\tau^2) \rho \frac{\lambda_{\max}(P)}{\lambda_{\min}(P)} \right) \leq \frac{\varepsilon}{2},$$

and to choose $\alpha \in \mathbb{R}_{>0}$ small enough such that $\alpha < \lambda_{\min}(P) \delta_\gamma$ and such that $x_t \in \Omega$ implies $k(x_t) \in \mathcal{U}$ and $|k(x_t)|^2 < \delta_\gamma$. Then, the condition $\dot{V}(x(t)) \leq -\varepsilon/2 \, |x(t)|^2$ holds for all $x_t \in \Omega$ whenever $V(x(t + \theta)) \leq \rho \, V(x(t))$ for all $\theta \in [-\tau, 0]$. Furthermore, the terminal region Ω, which was chosen as a sublevel set of the terminal cost functional $E(x_t)$, is positively invariant when applying the local control law $k(x_t)$.

In the second part, the functional

$$\tilde{E}(x_t) = \max_{\theta \in [-\tau, 0]} V(x(t + \theta)) \tag{4.32}$$

is considered for states $x_t \in \Omega$ inside the terminal region. Two cases can be distinguished:

$$(i) \ V(x(t)) < \frac{1}{\rho} \tilde{E}(x_t) \qquad \text{and} \qquad (ii) \ V(x(t)) \geq \frac{1}{\rho} \tilde{E}(x_t),$$

see also Figure 4.4.

For case (i), note that $V(x(t)) < \frac{1}{\rho} \tilde{E}(x_t)$ directly implies $V(x(t')) < \frac{1}{\rho} \tilde{E}(x_t)$ for all $t' \geq t$ because the Razumikhin-type condition is satisfied, i.e., $\dot{V} \leq 0$ whenever Condition (4.22) holds. Using

$$u(t')^T R u(t') \leq 2x(t')K^T R K x(t') + 2\lambda_{\max}(R) \, \tau^2 \|K_\tau\|_\tau^2 \|x_{t'}\|_\tau^2$$

in which $\|K_\tau\|_\tau = \sup_{\theta\in[-\tau,0]} \|K_\tau(\theta)\|$, it results that

$$\int_t^{t+\delta} F(x(t'), k(x_{t'}))\, dt' \leq \delta\, \Psi\, \tilde{E}(x_t) \quad \text{with } \Psi = \frac{\lambda_{\max}(Q + 2K^T RK) + 2\rho\tau^2 \|K_\tau\|_\tau^2 \lambda_{\max}(R)}{\rho\,\lambda_{\min}(P)}.$$

On the other hand, $\tilde{E}(x_{t+\delta}) < \frac{1}{\rho}\tilde{E}(x_t)$, which can be directly rewritten in the form $\tilde{E}(x_{t+\delta}) - \tilde{E}(x_t) \leq -\frac{\rho-1}{\rho}\tilde{E}(x_t)$. Choosing $E(x_t) = \beta\,\tilde{E}(x_t)$ with $\beta > \beta_{(i)} = \delta\,\Psi\,\frac{\rho}{\rho-1}$ guarantees that

$$E(x_{t+\delta}) - E(x_t) \leq -\int_t^{t+\delta} F(x(t'), k(x_{t'}))\, dt'. \tag{4.33}$$

For case (ii), it is assumed that $V(x(t)) \geq \frac{1}{\rho}\tilde{E}(x_t)$. For this case, again two subcases can be distinguished:

$$(iia)\ V(x(t+\delta-\tau)) < \frac{1}{\rho}\tilde{E}(x_t) \qquad \text{and} \qquad (iib)\ V(x(t+\delta-\tau)) \geq \frac{1}{\rho}\tilde{E}(x_t).$$

If $V(x(t+\delta-\tau)) < \frac{1}{\rho}\tilde{E}(x_t)$, then $\tilde{E}(x_{t+\delta}) < \frac{1}{\rho}\tilde{E}(x_t)$. Using arguments analogue to case (i) with replacing $V(x(t')) < \frac{1}{\rho}\tilde{E}(x_t)$ by $V(x(t')) < \tilde{E}(x_t)$ for all $t' \geq t$ shows that (4.33) holds for

$$\beta > \beta_{(iia)} = \rho\,\beta_{(i)} = \rho\,\delta\,\Psi\,\frac{\rho}{\rho-1} > \beta_{(i)}.$$

The last case to be considered is (iib), see also Figure 4.4. By the definition of $\tilde{E}$ it is clear that $\tilde{E}(x_{t+\delta}) \geq V(x(t+\delta-\tau))$. Combining this observation with $V(x(t+\delta-\tau)) \geq \frac{1}{\rho}\tilde{E}(x_t)$ and the fact that $\dot{V} \leq 0$ whenever $V(x(t')) \geq \frac{1}{\rho}\tilde{E}(x_{t'})$, we can directly obtain that $\tilde{E}(x_{t+\delta}) = V(x(t+\delta-\tau))$, see Figure 4.4. Note further that $\frac{1}{\rho}\tilde{E}(x_t) \leq V(x(t')) \leq \tilde{E}(x_t)$ and $|x(t')|^2 \geq \frac{1}{\rho\,\lambda_{\max}(P)}\tilde{E}(x_t)$ for all $t' \in [t, t+\delta-\tau]$. Hence, $\dot{V}(x(t')) \leq -\varepsilon/2\,|x(t')|^2$ for all $t' \in [t, t+\delta-\tau]$ and

$$\tilde{E}(x_{t+\delta}) - \tilde{E}(x_t) = V(x(t+\delta-\tau)) - \tilde{E}(x_t) \leq -\frac{\varepsilon}{2}\frac{\delta-\tau}{\rho\,\lambda_{\max}(P)}\tilde{E}(x_t).$$

Choosing $E(x_t) = \beta\,\tilde{E}(x_t)$ with

$$\beta > \beta_{(iib)} = \Psi\,\frac{\delta}{\delta-\tau}\frac{2}{\varepsilon}\rho\,\lambda_{\max}(P)$$

in which Ψ is defined as in case (i) guarantees again that (4.33) holds.

Hence, combining the results of cases (i), (iia) and (iib) implies that choosing $E(x_t) = \beta\,\tilde{E}(x_t)$ with

$$\beta > \max\left\{\beta_{(i)}, \beta_{(iia)}, \beta_{(iib)}\right\} = \max\left\{\beta_{(iia)}, \beta_{(iib)}\right\}$$

guarantees that Condition (4.33) is satisfied.

Therefore, the local control law $k(x_t)$ satisfies Condition (4.29) in Assumption 4.8. Furthermore, the terminal region Ω is positively invariant when applying $k(x_t)$ as shown in the first part of the proof. Assumption 4.4 is directly satisfied due to the definitions of F, E, and Ω, respectively. Finally, all assumptions necessary for Theorem 4.16 are satisfied. Hence, asymptotic stability is guaranteed. $\qquad\square$

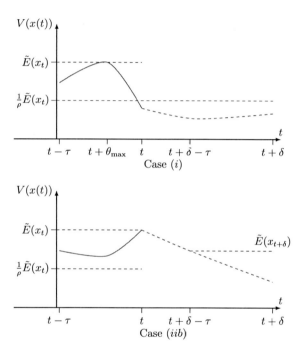

Figure 4.4: Sketch for proof of Theorem 4.17.

Example 4.18. *We consider again Example 4.6. As shown in Example 4.12, Assumption 4.7 is satisfied for the stage cost $F(x, u) = x^2 + 0.5u^2$, the Lyapunov-Razumikhin function $V(x) = x^2$, and the locally stabilizing control law $u(t) = k(x_t) = -2x(t)$. Furthermore, the nonlinear time-delay system (4.11) satisfies the Razumikhin-type condition (4.25) for all x_t inside the terminal region (4.31) with $\alpha = 1/4$. For cases (i) and (iia), we can use $\tilde{E}(x_{t+\delta}) - \tilde{E}(x_t) \leq -\frac{1}{2} E(x_t)$, whereas in case (ii) we obtain $\tilde{E}(x_{t+\delta}) - \tilde{E}(x_t) \leq -\frac{3(\delta-\tau)}{2} E(x_t)$. In all cases,*

$$\int\limits_{t}^{t+\delta} F(x(t'), k(x_{t'}))\, dt' = \int\limits_{t}^{t+\delta} 3x(t')^2\, dt' \leq \frac{3\delta}{2} \tilde{E}(x_t).$$

Hence, the terminal cost functional $E(x_t)$ defined in (4.30) satisfies Condition (4.29) for $\beta > \max\left\{3\delta, \frac{\delta}{\delta-\tau}\right\}$. Furthermore, the terminal region Ω and the terminal cost E satisfy Assumption 4.6. Hence, MPC according to Algorithm 4.4 with design parameters E and Ω asymptotically stabilizes the origin.

The scheme presented in this section allows to define the terminal region as a sublevel set of the terminal cost functional. However, two drawbacks have to be mentioned. First, the result requires the existence of a Lyapunov-Razumikhin function for the linearized closed-loop using the auxiliary local control law. This condition is more restrictive than only requiring stability as in Section 4.3.1, but similar to the conditions used in Section 4.3.2. The second, more severe drawback is that the sampling time of the MPC controller has to be chosen larger than the time-delay of the system. Clearly, this is problematic for systems which are open-loop unstable and/or exhibit large time-delays.

Design using LMIs

Similar to the results in the previous sections, we can also state an exemplary LMI condition for the design using Lyapunov-Razumikhin arguments.

Lemma 4.19 (LMI Condition for Assumption 4.7). *Consider the nonlinear time-delay system (4.1) and suppose that Assumptions 4.1–4.3 are satisfied. If there exist a symmetric matrix $\Lambda \succ 0$, a matrix Γ, and a constant positive scalar $\tilde{\varepsilon} \in \mathbb{R}_{>0}$ solving the following LMI*

$$\begin{bmatrix} \Xi_2 + \Lambda + \tilde{\varepsilon}I & A_\tau\Lambda \\ \star & -\frac{1}{\rho}\Lambda \end{bmatrix} \prec 0 \tag{4.34}$$

in which $\Xi_2 = \Lambda A^T + A\Lambda + \Gamma^T B^T + B\Gamma$, then Assumption 4.7 is satisfied with the Lyapunov-Razumikhin function $V(x) = x^T Px$ with $P = \Lambda^{-1}$, the local control law $u(t) = k(x_t) = Kx(t)$ with $K = \Gamma\Lambda^{-1}$, and $\varepsilon = \tilde{\varepsilon}\lambda_{\min}(P^2)$.

Proof. The proof is given in Appendix A.5. $\qquad\square$

Remark 4.20. *The application of LMIs similar to the ones used in Lemma 4.13 in order to obtain delay-dependent conditions is possible in principle. However, the proof is slightly more involved due to the model transformation to ξ coordinates, see the proof of Lemma 4.13 in Appendix A.3.*

4.3.4 Design with Lyapunov-Razumikhin and Exponential Weighting

In this section, we use Lyapunov-Razumikhin arguments in order to derive a novel terminal cost and a terminal region defined as a sublevel set of this terminal cost. In contrast to the previous result in Section 4.3.3, the assumption $\delta > \tau$ on the sampling time can be dropped due to an additional exponential weighting term in the terminal cost functional. This weighting term is similar to the one used in Sections 3.4 and 3.6. In contrast to the results presented previously, the weighting in this section is only used for the definition of the terminal cost.

The result can be summarized as follows.

Theorem 4.21 (Design using Lyapunov-Razumikhin and Exponential Weighting). *Consider the nonlinear time-delay system* (4.1) *and the quadratic stage cost* (4.10). *Suppose that Assumptions 4.1–4.3, and 4.7 are satisfied. Then, there exist a terminal cost functional of the form*

$$E(x_t) = \max_{\theta \in [-\tau, 0]} \beta(\theta) \cdot V(x(t + \theta)) \tag{4.35}$$

in which $\beta : [-\tau, 0] \to \mathbb{R}_{>0}$ *is defined by*

$$\beta(\theta) = \beta_0 \, e^{\mu \frac{\theta}{\tau}}, \quad \mu = \min \left\{ \ln(\rho), \frac{\tau \, \varepsilon}{2 \, \lambda_{\max}(P)} \right\} \in \mathbb{R}_{>0}, \quad \beta_0 \in \mathbb{R}_{>0},$$

and a terminal region

$$\Omega = \{ x_t \, : \, E(x_t) \leq \alpha \}, \tag{4.36}$$

in which $\alpha \in \mathbb{R}_{>0}$, *such that Assumptions 4.4 and 4.6 are satisfied. Furthermore, the closed-loop resulting from the application of the model predictive controller according to Algorithm 4.4 to system* (4.1) *is asymptotically stable.*

Proof. For $\gamma \in \mathbb{R}_{>0}$ chosen such that

$$\gamma \|P\| \left(4 + 2 \|K^T K\| + (1 + 2\tau^2 \|K_\tau\|_\tau^2) \, \rho \frac{\lambda_{\max}(P)}{\lambda_{\min}(P)} \right) \leq \frac{\varepsilon}{2},$$

we choose $\alpha \in \mathbb{R}_{>0}$ such that $\alpha < \beta_0 \, \lambda_{\min}(P) \, \delta_\gamma \, e^{-\mu}$ and such that $x_t \in \Omega$ implies $k(x_t) \in \mathcal{U}$ for all $\theta \in [-\tau, 0]$, which is always possible for small enough α. For this α, it was shown in Theorem 4.17 in Section 4.3.3 that the derivative of V along trajectories of the nonlinear system (3.1) satisfies the condition $\dot{V}(x(t)) \leq -\varepsilon/2 \, |x(t)|^2$ for all $x_t \in \Omega$ satisfying (4.22).

Define the auxiliary function $\bar{V}_t(\theta) = \frac{1}{\beta(\theta)} E(x_t)$, see Figure 4.5. Then it directly follows from the definition of E that $V(x(t + \theta)) \leq \bar{V}_t(\theta)$ for all $\theta \in [-\tau, 0]$. Moreover, due to the definition of μ, $\bar{V}_t(0) \geq \frac{1}{\rho} \bar{V}_t(\theta) \geq \frac{1}{\rho} V(x(t + \theta))$ for all $\theta \in [-\tau, 0]$ and

$$\dot{\bar{V}}_t(\theta) \geq -\frac{\varepsilon}{2 \, \lambda_{\max}(P)} \bar{V}_t(\theta) \geq -\frac{\varepsilon}{2} |x(t + \theta)|^2 \geq \dot{V}(x(t + \theta))$$

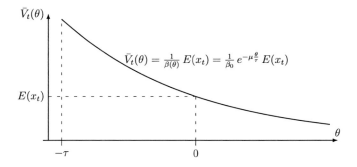

Figure 4.5: Upper bound $\bar{V}_t(\theta)$ on $V(x(t+\theta))$ used in the proof of Theorem 4.21.

whenever $V(x(t+\theta)) \geq \bar{V}_t(\theta)$. Hence, $V(x(t+\theta)) \leq \bar{V}_t(\theta)$ for all $\theta > -\tau$. Then, for all $\delta > 0$

$$E(x_{t+\delta}) = \max_{\theta \in [-\tau,0]} \beta(\theta) \cdot V(x(t+\delta+\theta)) \leq \max_{\theta \in [-\tau,0]} \beta(\theta) \cdot \bar{V}_t(\delta+\theta)$$

$$= \max_{\theta \in [-\tau,0]} \frac{\beta(\theta)}{\beta(\delta+\theta)} E(x_t) = e^{-\mu \frac{\delta}{\tau}} E(x_t).$$

Hence, by investigating $\delta \to 0$, we obtain

$$\dot{E}(x_t) \leq -\frac{\mu}{\tau} E(x_t). \tag{4.37}$$

In the next step, we show that $F(x(t),k(x_t)) \leq \frac{\psi}{\beta_0} E(x_t)$ for some constant $\psi \in \mathbb{R}_{>0}$. Using similar steps as in the proof of Theorem 4.17 in Section 4.3.3 with $\|K_\tau\|_\tau = \sup_{\theta \in [-\tau,0]} \|K_\tau(\theta)\|$, we obtain

$$F(x(t),k(x_t)) \leq \lambda_{\max}(Q+2K^TRK)|x(t)|^2 + 2\lambda_{\max}(R)\tau^2\|K_\tau\|_\tau^2\|x_t\|_\tau^2 \leq \frac{\psi}{\beta_0} E(x_t)$$

with $\psi = \frac{e^\mu}{\lambda_{\min}(P)} \bar{\psi}$ and $\bar{\psi} = \lambda_{\max}(Q+2K^TRK)+2\lambda_{\max}(R)\tau^2\|K_\tau\|_\tau^2$. By choosing $\beta_0 > \frac{\tau\psi}{\mu}$, it follows from (4.37) that $\dot{E}(x_t) \leq -F(x(t),k(x_t))$. By taking the definition of Ω as a sublevel set of E into account, positive invariance is ensured and, consequently, Assumption 4.6 is satisfied. The satisfaction of Assumption 4.4 is straightforward due to the definitions of F, E, and Ω, respectively. Asymptotic stability follows directly by use of Theorem 4.5, for which all necessary assumptions are satisfied. □

Example 4.22. *We consider again Example 4.6. As shown in Example 4.18, Assumption 4.7 is satisfied for the stage cost $F(x,u) = x^2 + 0.5u^2$, the Lyapunov-Razumikhin function $V(x) = x^2$, and the locally stabilizing control law $u(t) = k(x_t) = -2x(t)$. Moreover, it was shown that the nonlinear time-delay system (4.11) satisfies the Razumikhin-type condition (4.25) for all x_t inside the terminal region (4.31) with $\alpha = 1/4$. With*

$\mu = \min\{\ln(2), 3/2\} = \ln(2)$ *and* (4.37), *we directly obtain* $\dot{E}(x_t) \leq -\ln(2)\, E(x_t)$ *for the terminal cost functional* E *defined in* (4.35). *Furthermore, it is easy to see that* $\beta_0 F(x(t), k(x_t)) \leq 3E(x_t)$. *Hence,* $\dot{E}(x_t) \leq -F(x(t), k(x_t))$ *holds when* β_0 *is chosen such that* $\beta_0 > 3/\ln(2)$. *In this case, Assumption 4.6 is satisfied and MPC based on Algorithm 4.4 with terminal cost* E *and terminal region* Ω *asymptotically stabilizes the origin.*

As we will show in Section 4.4, it is possible to extend existing results on unconstrained MPC schemes, which rely on Ω defined as a sublevel set of E, to nonlinear time-delay systems using the design parameters E and Ω in Theorem 4.21.

4.3.5 Comparison of Different Approaches

We have presented four different design procedures for calculating the terminal cost functional and terminal region based on the Jacobi linearization. It is interesting to note that all four schemes contain the results of Chen and Allgöwer (1998) for delay-free systems as special case although the schemes rely on different assumptions and yield different results.

The main properties are summarized in Table 4.1. The first scheme in Section 4.3.1 can be considered as the most general scheme because it only requires the existence of a stabilizing linear control law and, in particular, does not require any Razumikhin-type condition. This can be beneficial because there does not exist any converse theorems regarding Lyapunov-Razumikhin functions, rendering Razumikhin-based approaches more restrictive (Gu et al., 2003). The main advantage of the three schemes using Lyapunov-Razumikhin arguments is that a simpler terminal region is obtained. As shown in Sections 4.3.3 and 4.3.4, it is also possible to define the terminal region as a sublevel set of the terminal cost with either a sampling time larger than the time-delay or an additional exponential weighting. This definition of the terminal region is not possible without the Razumikhin-type condition and is useful for the design of unconstrained MPC schemes as will be discussed in Section 4.4.

In Table 4.2, we summarize the results of the different design schemes for the system in Example 4.6, which was used throughout this section to illustrate the different results. Simulation results for this example will be given in Section 4.6.1.

4.4 Unconstrained MPC with Terminal Cost Functional

In this section, we consider an MPC setup in which the terminal constraint is omitted from the optimal control problem. However, the stability analysis still relies on such a terminal region and it is shown that for a defined set of initial states it is nevertheless possible to guarantee satisfaction of the terminal constraints without explicitly stating them in the optimization problem.

The results in this section extend the results derived by Limon et al. (2003, 2006) for discrete-time systems and by Graichen and Kugi (2010); Graichen et al. (2010) for continuous-time systems. A similar result using a modified terminal cost is reported by Hu and Linnemann (2002).

4.4.1 MPC Setup

We replace Problem 4.3 by the following optimization problem without terminal constraint, which has to be solved at each sampling time t_i given the measured state x_{t_i}.

Table 4.1: Comparison of the different design schemes based on the Jacobi linearization for calculating a suitable terminal region and terminal cost for model predictive control of nonlinear time-delay systems.

Design scheme	does not require Razumikhin function	modest complexity of terminal region	terminal region is sublevel set of terminal cost	no restrictions on sampling time
General linearization-based design (Section 4.3.1)	✓	–	–	✓
Combination of Lyapunov-Krasovskii and Lyapunov-Razumikhin (Section 4.3.2)	–	✓	–	✓
Design by Lyapunov-Razumikhin arguments without additional weighting (Section 4.3.3)	–	✓	✓	–
Design by Lyapunov-Razumikhin arguments with additional weighting (Section 4.3.4)	–	✓	✓	✓

Problem 4.23.

$$\underset{\bar{u}\in\mathcal{PC}([t_i,t_i+T],\mathbb{R}^m)}{\text{minimize}} \quad J_T(x_{t_i},\bar{u}) \tag{4.38a}$$

subject to

$$\dot{\bar{x}}(t';x_{t_i},t_i) = f(\bar{x}(t';x_{t_i},t_i),\bar{x}(t'-\tau;x_{t_i},t_i),\bar{u}(t')), \qquad t'\in[t_i,t_i+T], \tag{4.38b}$$

$$\bar{x}(t_i+\theta;x_{t_i},t_i) = x_{t_i}(\theta), \qquad \theta\in[-\tau,0], \tag{4.38c}$$

$$\bar{u}(t')\in\mathbb{U}, \qquad t'\in[t_i,t_i+T], \tag{4.38d}$$

in which

$$J_T(x_{t_i},\bar{u}) = \int_{t_i}^{t_i+T} F(\bar{x}(t';x_{t_i},t_i),\bar{u}(t'))\,dt' + E(\bar{x}_{t_i+T}).$$

We denote the optimal open-loop input trajectory by $u_T^*(t';x_{t_i},t_i)$ for all $t'\in[t_i,t_i+T]$. The associated optimal cost is denoted by $J_T^*(x_{t_i})$ and the associated predicted state trajectory is $x_{t'}^*$, $t'\in[t_i,t_i+T]$.

Table 4.2: Comparison of terminal cost and the terminal region based on the different design schemes for Example 4.6.

design scheme	terminal cost $E(x_t)$	terminal region $\Omega = \{x_t \in \mathcal{C}_\tau : \ldots\}$
Section 4.3.1 (Example 4.9, page 75)	$x(t)^2 + \int\limits_{-\tau}^{0} x(t+\theta)^2 \, d\theta$	$E(x_t) \leq \frac{1}{4}, \|x_t\|_\tau \leq \frac{1}{2}$
Section 4.3.2 (Example 4.12, page 77)	$x(t)^2 + \int\limits_{-\tau}^{0} x(t+\theta)^2 \, d\theta$	$\max\limits_{\theta \in [-\tau,0]} x(t+\theta)^2 \leq \frac{1}{4}$
Section 4.3.3 (Example 4.18, page 84)	$\beta \max\limits_{\theta \in [-\tau,0]} x(t+\theta)^2$	$\max\limits_{\theta \in [-\tau,0]} x(t+\theta)^2 \leq \frac{1}{4}$
Section 4.3.4 (Example 4.22, page 86)	$\max\limits_{\theta \in [-\tau,0]} \frac{3}{\ln(2)} 2^\theta x(t+\theta)^2$	$E(x_t) \leq \frac{3}{4\ln(2)}$

In comparison to Problem 4.3, only the terminal constraint (4.2e) is removed from the finite horizon optimal control problem. However, our analysis relies on the fact that for a defined set of initial states the terminal constraint is nevertheless guaranteed to be satisfied. More precisely, a sublevel set of the optimal cost belongs to the region of attraction of the origin of the closed-loop even if no terminal constraint is included in the optimal control problem. These results extends the results by Graichen and Kugi (2010); Limon et al. (2006) to nonlinear time-delay systems.

As before, the control input to the system is defined in the usual receding horizon fashion as stated in the following algorithm.

Algorithm 4.24 (Model Predictive Control for Nonlinear Time-Delay Systems without Terminal Constraint). *At each sampling instant $t_i = i\delta$, $i \in \mathbb{N}_0$, measure the state x_{t_i} and solve Problem 4.23. Apply the input*

$$u_{\mathrm{MPC}}(t) = u_T^*(t; x_{t_i}, t_i), \quad t_i \leq t < t_i + \delta. \tag{4.39}$$

to the system until the next sampling instant $t_{i+1} = t_i + \delta$.

4.4.2 Asymptotic Stability

For the stability analysis of this MPC scheme, we require the following assumptions, which are essentially identical to Assumptions 4.4 and 4.6.

Assumption 4.10. *The terminal cost functional $E : \mathcal{C}_\tau \to \mathbb{R}_{\geq 0}$ is continuously differentiable, positive definite, and there exist class $\mathcal{K}_\infty$ functions $\underline{\alpha}_E, \overline{\alpha}_E : \mathbb{R}_{\geq 0} \to \mathbb{R}_{\geq 0}$ such that $\underline{\alpha}_E(|x(t)|) \leq E(x_t) \leq \overline{\alpha}_E(\|x_t\|_\tau)$. The stage cost $F : \mathbb{R}^n \times \mathbb{U} \to \mathbb{R}_{\geq 0}$ is continuous, $F(0,0) = 0$, and there is a class $\mathcal{K}_\infty$ function $\underline{\alpha}_F : \mathbb{R}_{\geq 0} \to \mathbb{R}_{\geq 0}$ such that*

$$F(x,u) \geq \underline{\alpha}_F(|x|) \quad \text{for all } x \in \mathbb{R}^n, u \in \mathbb{U}. \tag{4.40}$$

Assumption 4.11. *There is a constant $\alpha \in \mathbb{R}_{>0}$ defining the set*

$$\Omega = \{x_t : E(x_t) \leq \alpha\} \subseteq \mathcal{C}_\tau$$

and there exists a locally asymptotically stabilizing controller $u(t) = k(x_t) \in \mathbb{U}$ for the nonlinear time-delay system (4.1) such that the set Ω is controlled positively invariant and

$$\forall x_t \in \Omega \;:\; \dot{E}(x_t) \leq -F(x(t), k(x_t)). \tag{4.41}$$

In order to obtain a suitable lower bound on the cost for states outside the terminal region, see Lemma 4.26, we require in addition the following assumptions.

Assumption 4.12. *The prediction horizon is chosen such that $T > 2\tau$. There exists a quadratic lower bound on the stage cost, i.e., there is $\lambda_F \in \mathbb{R}_{>0}$ such that $F(x, u) \geq \lambda_F |x|^2$ for all $x \in \mathbb{R}^n$, $u \in \mathbb{U}$. Furthermore, the function $f : \mathbb{R}^n \times \mathbb{R}^n \times \mathbb{R}^m \to \mathbb{R}^n$ is globally Lipschitz continuous.*

For our stability analysis, we first state two intermediate results in the following lemmata. The first lemma is based on (Limon et al., 2006, Lemma 1) and states that if the predicted terminal state is not contained in the terminal region Ω, then every state along the predicted trajectory within the prediction horizon is outside of Ω.

Lemma 4.25. *Consider the optimal control problem 4.23 and suppose that Assumption 4.11 is satisfied. If $x_{t_i+T}^* \notin \Omega$, then $x_{t'}^* \notin \Omega$ for all $t' \in [t_i, t_i + T]$.*

Proof. For the sake of contradiction, assume that there exists a $t^\dagger \in [0, T]$ such that $x_{t_i+t^\dagger}^* \in \Omega$ and that $x_{t_i+T}^* \notin \Omega$ or, equivalently, $E(x_{t_i+T}^*) > \alpha$. By the principle of optimality (Bellman, 1957, Chapter III, §3), we know that $x_{t'}^*$ is an optimal endpiece on the interval $t' \in [t_i + t^\dagger, t_i + T]$. Moreover, $J_{T-t^\dagger}^*(x_t) \leq E(x_t)$ holds for all $x_t \in \Omega$ because of Condition (4.41), see also Mayne et al. (2000) for the equivalent result in discrete-time. Hence, we know that

$$E(x_{t_i+t^\dagger}^*) \geq J_{T-t^\dagger}^*(x_{t_i+t^\dagger}^*) \geq E(x_{t_i+T}^*) > \alpha.$$

Therefore, it follows that $E(x_{t_i+t^\dagger}^*) > \alpha$ and $x_{t_i+t^\dagger}^* \notin \Omega$, which contradicts the assumption. The proof of the lemma is complete. $\qquad\square$

Note that the proof of Lemma 4.25 requires Ω to be defined as a sublevel set of the terminal cost E. This allows the use of the terminal cost and terminal region as defined in Sections 4.3.3 and 4.3.4. However, it is not possible to use the results from Sections 4.3.1 and 4.3.2.

In (Limon et al., 2006, Assumption 2), it was required that there exists $d \in \mathbb{R}_{>0}$ such that $F(x, u) > d$ holds for all $x \notin \Omega$ and all $u \in \mathbb{U}$. While this is a reasonable assumption for delay-free systems, this clearly does not hold true for the problem setup considered in this chapter and almost all other results on MPC for time-delay systems in the literature, with a notable exception presented by Brunner (2010). Note that the stage cost F only penalizes the instantaneous state $x(t)$ and not the full state x_t of the system. However, a weaker property is sufficient for our stability analysis. Instead of each single time instant, we can consider the whole prediction horizon in order to derive an appropriate lower bound. By the definition of the terminal region as a sublevel set, the bound $E(x_{t'}) > \alpha$ holds for all $x_{t'} \notin \Omega$. From this, a lower bound on the integral $\int_{t'-2\tau}^{t'} |x(t)|^2 dt$ can be derived, which implies the existence of $\hat{d} \in \mathbb{R}_{>0}$ such that $\int_{t'-2\tau}^{t'} F(x,u)(t)dt \geq \hat{d}$ for all $t' \in [t+2\tau, T]$. From this, the inequality

$$\int_t^{t+T} F(x(\theta), u(\theta)) d\theta \geq \hat{d} \text{ floor}(T/(2\tau)) \geq \hat{d} \, (T/(2\tau) - 1)$$

follows. These results are summarized in Lemma 4.26.

Lemma 4.26. *Consider the cost functional defined in Problem 4.23 and suppose that Assumptions 4.1–4.3 and 4.10–4.12 are satisfied. Furthermore, suppose that $x_{t'}^* \notin \Omega$ for all $t' \in [t_i, t_i + T]$ in which $\Omega = \{x_t : E(x_t) \leq \alpha\}$. Then, there exists a $\hat{d} \in \mathbb{R}_{>0}$ such that*

$$\int_{t_i}^{t_i+T} F(x(t'), u(t')) \, dt' > \hat{d} \, (T/(2\tau) - 1) \, .$$

Proof. Define two auxiliary functions $y, z : [-\tau, 0] \to \mathbb{R}_{\geq 0}$ by $y(\theta) = |x(t' + \theta - \tau)|$ and $z(\theta) = |x(t' + \theta)|$, which describe $|x(t)|$ in the intervals $t \in [t' - 2\tau, t' - \tau]$ and $t \in [t' - \tau, t']$, respectively. In the following, we consider an arbitrary, but fixed, time instant $t' \in [t_i + \tau, t_i + T]$ and derive a lower bound on $\int_{t'-2\tau}^{t'} F(x(t), u(t)) \, dt$. The assumption $x_{t'}^* \notin \Omega$ is equivalent to $E(x_{t'}) > \alpha$ and, consequently,

$$\|x_{t'}\|_\tau > \alpha_1 \, , \qquad \text{in which} \quad \alpha_1 = \overline{\alpha}_E^{-1}(\alpha) \, .$$

The integral $\int_{t'-\tau}^{t'} F(x(t), u(t)) dt$ will be minimal if $z(\theta)$ reaches its maximum

$$\max_{\theta \in [-\tau, 0]} z(\theta) = \|x_{t'}\|_\tau > \alpha_1$$

at some time $\tilde{\theta} = \arg\max_{\theta \in [-\tau, 0]} z(\theta) \in [-\tau, 0]$ and descends in both positive and negative time direction with the largest possible gradient. Note that we do not make any assumptions on a particular shape of the function $y(\theta)$. Without loss of generality, it can be assumed that $\tilde{\theta} = -\tau$. In the following, a rectangle of maximum width $\tau/2$ will be used in order to obtain a lower bound for the integral $\int_{t'-\tau}^{t'} F(x(t), u(t))dt$. Therefore, analogue arguments also hold if the maximum is not attained at $\tilde{\theta} = -\tau$, but at another time, see Figure 4.6.

Two cases can now be distinguished:

$$(i) \int_{t'-2\tau}^{t'} |x(t)|^2 \, dt > \alpha_1^2 \tau/8 \qquad \text{and} \qquad (ii) \int_{t'-2\tau}^{t'} |x(t)|^2 \, dt \leq \alpha_1^2 \tau/8 \, .$$

For case (i), we can directly use the lower bound

$$\int_{t'-2\tau}^{t'} F(x(t), u(t)) \, dt > \lambda_F \, \alpha_1^2 \tau/8 \, .$$

For case (ii), we derive a lower bound on $z(\theta)$, which in turn gives a time θ^* such that $z(\theta) \geq \alpha_1/2$ for all $\theta \in [\tilde{\theta}, \theta^*]$, see Figure 4.6. To this end, note that due to Assumptions 4.3

and 4.12, the function f is globally Lipschitz continuous and u is bounded. Hence, there exist constants $L_0, L_1, L_2 \in \mathbb{R}_{>0}$ such that

$$\forall x(t), x(t-\tau) \in \mathbb{R}^n,\ u \in \mathbb{U}\ :\ |f(x(t), x(t-\tau), u(t))| \leq L_0 + L_1|x(t)| + L_2|x(t-\tau)|.$$

Hence, we have the following lower bound on $z(\theta)$

$$z(\theta) \geq \bar{z}(\theta) = \max\left(0, \alpha_1 - \underbrace{(L_0 + L_1\alpha_1)}_{=L}(\theta + \tau) - L_2 \cdot \int_{-\tau}^{\theta} y(\theta')\, d\theta'\right).$$

Note that $\bar{z}(\theta)$ is strictly monotonically decreasing as long as $\bar{z}(\theta) > 0$. Thus, θ^* satisfying

$$\alpha_1/2 - L\left(\theta^* + \tau\right) - L_2 \cdot \int_{-\tau}^{\theta^*} y(\theta')\, d\theta' \geq 0 \tag{4.42}$$

guarantees that $z(\theta) \geq \alpha_1/2$ for all $\theta \in [0, \theta^*]$ as desired. The well-known relation

$$\left(\frac{1}{\theta_2 - \theta_1} \int_{\theta_1}^{\theta_2} y(\theta')\, d\theta'\right)^2 \leq \frac{1}{\theta_2 - \theta_1} \int_{\theta_1}^{\theta_2} y^2(\theta')\, d\theta',$$

which follows from the Cauchy-Schwarz Inequality (Bronstein et al., 2000), and the fact that $\int_{t'-2\tau}^{t'} |x(t+\theta)|^2\, dt \leq \alpha_1^2 \tau/8$ finally imply

$$\int_{-\tau}^{\theta^*} y(\theta')\, d\theta' \leq \sqrt{(\theta^* + \tau)\alpha_1^2\tau/8}.$$

Hence, the solution of

$$\alpha_1/2 - L\left(\theta^* + \tau\right) - L_2\sqrt{(\theta^* + \tau)\alpha_1^2\tau/8} = 0,$$

which can be directly calculated as

$$\theta^* + \tau = \theta^* - \tilde{\theta} = \Delta\theta = L_2^2/(4L^2)\left(-\sqrt{\tau/8}\alpha_1 + \sqrt{\alpha_1^2\tau/8 + 2L\alpha_1/L_2^2}\right)^2,$$

satisfies (4.42). Consequently, it is possible to establish the following lower bound

$$\int_{t'-2\tau}^{t'} F(x(t), u(t))\, dt > \lambda_F \min\left\{\alpha_1^2\tau/8, \alpha_1^2\Delta\theta/4\right\} = \hat{d} > 0,$$

which covers both cases (i) and (ii). Also note that $\Delta\theta \leq \tau/2$ is implicitly included. It follows that if $x_{t'}^* \notin \Omega$ for all $t' \in [t_i, t_i + T]$, then

$$\int_{t_i}^{t_i+T} F(x(t'), u(t'))\, dt' > \hat{d}\ \mathrm{floor}(T/(2\tau)) \geq \hat{d}\left(T/(2\tau) - 1\right).$$

This completes the proof. $\qquad\qquad\qquad\qquad\qquad\qquad\qquad\qquad\qquad\qquad\qquad\square$

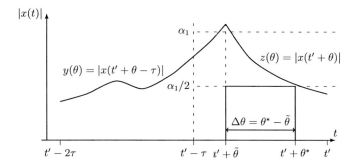

Figure 4.6: Sketch for proof of Lemma 4.26.

Using the lower bound derived in the previous lemma, we can formulate the main result of this section as stated in the following theorem.

Theorem 4.27 (Stability of Unconstrained MPC with Terminal Cost Functional for Nonlinear Time-Delay Systems). *Consider the nonlinear time-delay system* (4.1) *and suppose that Assumptions 4.1–4.3 and 4.10–4.12 are satisfied and that the prediction horizon satisfies* $T > 2\tau$. *Then, the closed-loop system resulting from the application of the model predictive controller according to Algorithm 4.24 to system* (4.1) *is asymptotically stable. A subset of the region of attraction is given by*

$$\Gamma_T = \left\{ x_t \in \mathcal{C}_\tau \ : \ J_T^*(x_t) \leq \hat{d}\left(T/(2\tau) - 1\right) + \alpha \right\}. \tag{4.43}$$

Proof. The proof is similar to the proof of (Limon et al., 2006, Theorem 1). Note that the region Ω is contained in Γ_T because $J_T^*(x_t) \leq E(x_t) \leq \alpha < \hat{d}\left(T/(2\tau) - 1\right) + \alpha$ for all $x_t \in \Omega$.

First, it is proven by contradiction that for any $x_{t_i} \in \Gamma_T$, the optimal solution satisfies the terminal constraint. For the sake of contradiction, assume $x_{t_i} \in \Gamma_T$ and that the terminal constraint is not satisfied. From Lemma 4.25, it can be inferred that if the optimal trajectory is such that the terminal region is not reached at the end of the prediction horizon, then all states along this trajectory are outside of Ω. Hence, by Lemma 4.26,

$$J_T^*(x_{t_i}) > \hat{d}\left(T/(2\tau) - 1\right) + \alpha$$

whenever $T > 2\tau$. But this contradicts the assumption $x_{t_i} \in \Gamma_T$ with Γ_T defined in (4.43). Therefore, we have shown that for all $x_{t_i} \in \Gamma_T$ the optimal solution of Problem 4.23 satisfies the terminal constraint $\bar{x}_{t_i+T} \in \Omega$ although it is not explicitly included in the optimization problem.

Second, it is proven that Γ_T is a positively invariant set for the closed-loop system resulting from the application of the model predictive controller according to Algorithm 4.24 to system (4.1). Suppose that $x_{t_i} \in \Gamma_T$. Since the terminal constraint is satisfied and Assumption 4.11 holds, the assertion of Lemma A.5 in Appendix A.1 holds. By using (A.4) and the non-negativity of the stage cost F, it is directly clear that for all $t^* \in [t_i, t_i + \delta]$

$$J_T^*(x_{t^*}) \leq J_T^*(x_{t_i}) - \int_{t_i}^{t^*} F(x(t'), u(t')) \, dt' \leq \hat{d} \, (T/(2\tau) - 1) + \alpha \,, \qquad (4.44)$$

which implies that $x_{t^*} \in \Gamma_T$ and, in particular, $x_{t_{i+1}} \in \Gamma_T$. Iterative application of these arguments shows positive invariance of Γ_T for all times.

Equation (4.44) shows that the optimal cost is non-increasing along trajectories of the closed-loop and that the only trajectory for which the optimal cost is not decreasing is the steady state $x_{t,s} = 0$. Therefore, stability and asymptotic stability can be proven along the lines of the proof of Theorem 4.5 in Appendix A.1. □

4.4.3 Summary

In this section, we have presented stability conditions for an MPC scheme without terminal constraints for nonlinear time-delay systems. The omission of the terminal constraints makes the approach computationally more attractive than other schemes with terminal constraints. However, the region of attraction might be smaller than for MPC with terminal constraints, but in most cases significantly larger than the estimate (4.43), which is based on a general, yet conservative analysis. Further results and extensions can be found in Brunner (2010), which proposes different alternative MPC formulations, e.g., similar to the result of Hu and Linnemann (2002) for delay-free systems.

4.5 Unconstrained MPC without Terminal Cost Functional

There exist several MPC schemes for nonlinear time-delay systems which guarantee closed-loop stability. These schemes can be roughly categorized into schemes with terminal constraints, see (Angrick, 2007; Raff et al., 2007) and Section 4.2, and unconstrained MPC schemes which use additional terminal weighting functionals, see (Brunner, 2010; Kwon et al., 2001a,b; Lu, 2011; Mahboobi Esfanjani and Nikravesh, 2009a, 2011) and Section 4.4. All of these schemes require a positively invariant terminal region. Furthermore, with the exception of Angrick (2007); Raff et al. (2007) who use an extended zero terminal state constraint, a control Lyapunov-Krasovskii functional is used as a terminal cost functional in the MPC setup. Calculating a control Lyapunov-Krasovskii functional for nonlinear time-delay systems is in general a difficult task. Even if a control Lyapunov-Krasovskii functional is known for the Jacobi linearization of the system about the origin, which by itself is not simple, it is a non-trivial problem to obtain an appropriate terminal cost functional and invariant terminal region for the nonlinear system, see Section 4.3. All schemes for calculating these stabilizing design parameters, see the schemes derived in Section 4.3 and the other aforementioned references, either require restrictive Razumikhin conditions or yield complicated terminal regions and/or terminal cost functionals, which are unattractive for an online implementation.

In order to overcome these difficulties, we consider MPC with finite horizon cost functionals containing neither terminal constraints nor terminal penalty terms in this section. The results extend the results on unconstrained MPC for continuous-time systems in Section 3.2 and the results concerning discrete-time systems (Grimm et al., 2005; Grüne, 2009; Grüne et al., 2010a). Note that the latter results hold, at least in principle, also for infinite-dimensional systems. For instance, the aforementioned results have been exemplarily applied to certain classes of partial differential equations, see Altmüller et al. (2010a,b, 2012). However, these results cannot be directly transferred to the MPC setup for time-delay systems considered in this chapter, as well as the other existing MPC schemes for time-delay systems, see the references in Section 2.2.3. The main reason is that the stage cost is not positive definite in the full state, but only penalizes the instantaneous state of the system. Hence, additional arguments are required in order to guarantee closed-loop stability. First, we introduce a modified controllability assumption which extends Assumption 3.5 in a suitable way towards time-delay systems. Based on this assumption, we derive conditions on the prediction horizon to guarantee stability of the closed-loop. It is particularly interesting to note that in contrast to essentially all other MPC schemes with guaranteed stability, the optimal cost is *not* used as Lyapunov function(al) of the closed-loop, and indeed the optimal cost can increase along trajectories of the closed-loop due to the influence of the delayed states. However, stability is guaranteed because the infinite horizon cost is bounded by a function of the initial state and the finite horizon optimal cost at initial time. Furthermore, the unconstrained MPC setup allows to make statements concerning performance of the closed-loop.

4.5.1 MPC Setup

With respect to performance, the goal is to minimize the infinite horizon cost functional $J_\infty(\varphi, u)$

$$\underset{\bar{u} \in \mathcal{PC}(\mathbb{R}_{\geq 0}, \mathbb{U})}{\text{minimize}} \ J_\infty(\varphi, u)\,, \quad \text{with } J_\infty(\varphi, u) = \int_0^\infty F(x(t), u(t))\, dt\,,$$

subject to the system dynamics (4.1). The associated optimal cost of this infinite horizon optimal control problem is denoted by $J_\infty^\star(\varphi) = \underset{\bar{u} \in \mathcal{PC}(\mathbb{R}_{\geq 0}, \mathbb{U})}{\min} J_\infty(\varphi, u)$ in this section and we use the following condition concerning the stage cost F analogue to Assumptions 3.4 and 4.4.

Assumption 4.13. *The stage cost* $F : \mathbb{R}^n \times \mathbb{U} \to \mathbb{R}_{\geq 0}$ *is continuous,* $F(0,0) = 0$, *and there exists a class* $\mathcal{K}_\infty$ *function* $\underline{\alpha}_F : \mathbb{R}_{\geq 0} \to \mathbb{R}_{\geq 0}$ *such that*

$$F(x, u) \geq F(x, 0) \geq \underline{\alpha}_F(|x|) \quad \text{for all } x \in \mathbb{R}^n,\, u \in \mathbb{U}\,. \tag{4.45}$$

In addition, we require the following assumption.

Assumption 4.14. *The prediction horizon* T *is chosen such that* $T > \tau + \delta$.

Assumption 4.14 is essentially only needed due to technical reasons in the proof of Lemma 4.32. In general, this assumption is not restrictive, in particular when considering small sampling times. In most cases, it is desirable to choose the prediction horizon larger than the time-delay or this might even be required in the case of using terminal constraints.

Remark 4.28. *The results of Section 3.2 on unconstrained MPC for nonlinear systems are not applicable in the setup presented in this section because the stage cost F is not positive definite in the full state x_t. Thus, (3.2) in Assumption 3.4 is not satisfied and we cannot expect that Assumption 3.5 is satisfied for a nonlinear time-delay system.*

Since infinite horizon problems are often computationally intractable, we use the finite horizon cost functional

$$J_T(\varphi, u) = \int_0^T F(x(t), u(t)) \, dt \,,$$

in which T is the prediction horizon, analogue to Section 3.2. The open-loop finite horizon optimal control problem at sampling time t_i given the measured state x_{t_i} is now formulated as follows.

Problem 4.29.

$$\underset{\bar{u} \in \mathcal{PC}([t_i, t_i + T], \mathbb{R}^m)}{\text{minimize}} \quad J_T(x_{t_i}, \bar{u}) \tag{4.46a}$$

subject to

$$\dot{\bar{x}}(t'; x_{t_i}, t_i) = f(\bar{x}(t'; x_{t_i}, t_i), \bar{x}(t' - \tau; x_{t_i}, t_i), \bar{u}(t')) \,, \qquad t' \in [t_i, t_i + T] \,, \tag{4.46b}$$

$$\bar{x}(t_i + \theta; x_{t_i}, t_i) = x(t_i + \theta) \,, \qquad \theta \in [-\tau, 0] \,, \tag{4.46c}$$

$$\bar{u}(t') \in \mathbb{U} \,, \qquad t' \in [t_i, t_i + T] \,, \tag{4.46d}$$

in which

$$J_T(x_{t_i}, \bar{u}) = \int_{t_i}^{t_i + T} F(\bar{x}(t'; x_{t_i}, t_i), \bar{u}(t')) \, dt' \,.$$

In Problem 4.29, $\bar{x}(t'; x_{t_i}, t_i)$ denotes the predicted trajectory starting from initial condition $\bar{x}(t_i + \theta; x_{t_i}, t_i) = x(t_i + \theta)$, $\theta \in [-\tau, 0]$ and driven by $\bar{u}(t')$ for $t' \in [t_i, t_i + T]$. We assume that the optimal open-loop control which minimizes $J_T(x_{t_i}, \bar{u})$ is given by $u_T^*(t'; x_{t_i}, t_i)$ for all $t' \in [t_i, t_i + T]$. The associated optimal cost is denoted by $J_T^*(x_{t_i})$ and the associated predicted trajectory is $x_T^*(t'; x_{t_i}, t_i)$, $t' \in [t_i, t_i + T]$. For given sampling time δ with $0 < \delta \le T$, the control input to the system is defined by the following algorithm in the usual receding horizon fashion analogue to Algorithm 3.6.

Algorithm 4.30 (Unconstrained Model Predictive Control for Nonlinear Time-Delay Systems). *At each sampling instant $t_i = i\delta$, $i \in \mathbb{N}_0$, measure the state x_{t_i} and solve Problem 4.29. Apply the input*

$$u_{\text{MPC}}(t) = u_T^*(t; x_{t_i}, t_i) \,, \quad t_i \le t < t_i + \delta \,. \tag{4.47}$$

to the system until the next sampling instant $t_{i+1} = t_i + \delta$.

Note that in this section we consider MPC with a finite horizon optimal control problem containing neither terminal constraints nor terminal penalty terms in the cost functional. Thus, we do not require to calculate a local control Lyapunov-Krasovskii functional for the system in a region around the origin. This is in contrast to most of the previous work on MPC for nonlinear time-delay systems, see (Angrick, 2007; Kwon et al., 2001a,b; Mahboobi Esfanjani and Nikravesh, 2009a; Raff et al., 2007) and Section 4.2. Instead, we use a less restrictive controllability assumption along the lines of the work presented by Grimm et al. (2005); Grüne (2009); Grüne et al. (2010a) for discrete-time systems, which we also discussed in Chapter 3 for continuous-time delay-free systems.

4.5.2 Controllability Assumption and Implications

In the following, we introduce an extended version of the Controllability Assumption 3.5 which is appropriate for the nonlinear time-delay systems considered in this chapter.

Assumption 4.15 (Controllability Assumption for Time-Delay Systems). *For all* $T' \in \mathbb{R}_{\geq 0}$ *and for all* $\varphi \in \mathcal{C}_\tau = \mathcal{C}([-\tau, 0], \mathbb{R}^n)$, *there exists a piece-wise continuous input trajectory* $\hat{u}(\cdot; \varphi, 0)$ *with* $\hat{u}(t; \varphi, 0) \in \mathbb{U}$ *for all* $t \in [0, T']$ *and*

$$J_{T'}^*(\varphi) \leq J_{T'}(\varphi, \hat{u}) \leq B(T') \left(F(\varphi(0), 0) + \int_{-\tau}^{0} F(\varphi(t'), 0) dt' \right), \tag{4.48}$$

in which $B : \mathbb{R}_{\geq 0} \to \mathbb{R}_{>0}$ *is a continuous, non-decreasing, and bounded function.*

Note that Assumption 4.15 is a natural extension of Assumption 3.5, which is directly recovered for $\tau = 0$. Also note that the candidate input trajectory $\hat{u}$ has to be feasible in the sense that it satisfies the input constraints, but it is not required to be optimal.

Without loss of generality, we consider the two consecutive sampling instants $t_0 = 0$ and $t_1 = \delta$ in the following. Since system (4.1) is time-invariant, all results hold analogously for any other two consecutive sampling instants t_i and t_{i+1}. Furthermore, we use the following abbreviations

$$F^*(t; t_i) = \begin{cases} F(x_T^*(t; x_{t_i}, t_i), u_T^*(t; x_{t_i}, t_i)), & t \in [t_i, t_i + T] \\ F(x(t), u(t)), & t \in [t_i - \tau, t_i[\end{cases}$$

for $t_i \in \{0, \delta\}$. From this definition, it directly follows that $J_T^*(x_{t_i}) = \int_{t_i}^{t_i+T} F^*(t'; t_i) \, dt'$.

In the following, we state two intermediate results in Lemmata 4.32 and 4.33, which are based upon the Controllability Assumption 4.15 and the auxiliary result of Lemma 4.31. Lemma 4.32 uses the optimality of $J_T^*(x_{t_i+\delta})$ in addition to the controllability assumption in order to derive an upper bound on $J_T^*(x_{t_i+\delta})$ in terms of the endpiece of the predicted trajectory calculated at time t_i. In Figure 4.7, this can be interpreted as giving an upper bound on the cost of the blue dotted line in terms of the red loosely dashed line. Lemma 4.33 applies the principle of optimality (Bellman, 1957, Chapter III, §3), i.e., the trajectory $F^*(t; t_i)$ calculated at time t_i is an optimal endpiece on the interval $[t_i + \delta, t_i + T]$. Hence, it is also possible to derive an upper bound based on the controllability assumption. In Figure 4.7, the result can be interpreted as giving an upper bound on the cost of the red loosely dashed line in terms of the green dashed line and the black solid line, which accounts for the influences of the delayed states due to the time-delay τ. Also note the similarity of these results to the results of Lemmata 3.20 and 3.21 for delay-free systems.

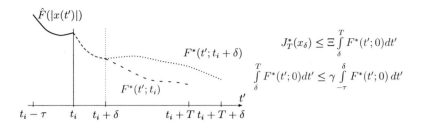

Figure 4.7: Sketch for Lemmata 4.32 and 4.33. The optimal trajectory calculated at time t_i is depicted by the (green and red) dashed line, the optimal trajectory calculated at $t_i + \delta$ is depicted by the (blue) dotted line.

Lemma 4.31 (Auxiliary Result for the Proofs of Lemmata 4.32 and 4.33). *For all $t_1, t_2 \in \mathbb{R}$ with $t_1 < t_2$, $\tau \in \mathbb{R}_{>0}$, and for any positive integrable function $F : [t_1 - \tau, t_2] \to \mathbb{R}_{\geq 0}$, the following holds*

$$\min_{t \in [t_1, t_2]} \left(F(t) + \int_{t-\tau}^{t} F(t') \, dt' \right) \leq \frac{1+\tau}{t_2 - t_1} \int_{t_1-\tau}^{t_2} F(t') \, dt' \,.$$

Proof. The proof is given in Appendix A.6. □

Lemma 4.32 (Calculation of Ξ). *Suppose that Assumptions 4.1–4.3, 4.13, 4.14, and Controllability Assumption 4.15 are satisfied for the nonlinear time-delay system (4.1). Then,*

$$J_T^*(x_\delta) \leq \Xi \int_{\delta}^{T} F^*(t'; 0) dt' \tag{4.49}$$

with $\Xi = 1 + B(T) \frac{1+\tau}{T-\tau-\delta}$.

Proof. Due to Controllability Assumption 4.15

$$J_T^*(x_\delta) \leq \int_{\delta}^{t} F^*(t'; 0) \, dt' + B(T + \delta - t) \left(F^*(t; 0) + \int_{t-\tau}^{t} F^*(t'; 0) \, dt' \right) \tag{4.50}$$

holds for all $t \in [\delta, T]$. Furthermore, we directly obtain the following relations

$$\int_{\delta}^{t} F^*(t'; 0) \, dt' \leq \int_{\delta}^{T} F^*(t'; 0) \, dt' \qquad \text{and} \qquad B(T + \delta - t) \leq B(T)$$

for all $t \in [\delta, T]$ due to non-negativity of F^* and the non-decreasing property of B.

Since (4.50) holds for all $t \in [\delta, T]$, it results

$$J_T^*(x_\delta) \leq \min_{t \in [\delta, T]} \left(\int_\delta^t F^*(t'; 0)\, dt' + B(T + \delta - t) \left(F^*(t; 0) + \int_{t-\tau}^t F^*(t'; 0)\, dt' \right) \right)$$

$$\leq \int_\delta^T F^*(t'; 0)\, dt' + B(T) \min_{t \in [\delta, T]} \left(F^*(t; 0) + \int_{t-\tau}^t F^*(t'; 0)\, dt' \right).$$

Using the auxiliary result stated in Lemma 4.31 yields

$$\min_{t \in [\delta, T]} \left(F^*(t; 0) + \int_{t-\tau}^t F^*(t'; 0)\, dt' \right) \leq \min_{t \in [\delta + \tau, T]} \left(F^*(t; 0) + \int_{t-\tau}^t F^*(t'; 0)\, dt' \right)$$

$$\leq \frac{1 + \tau}{T - \tau - \delta} \int_\delta^T F^*(t'; 0)\, dt',$$

and we finally arrive at

$$J_T^*(x_\delta) \leq \underbrace{\left(1 + B(T) \frac{1 + \tau}{T - \tau - \delta} \right)}_{= \Xi} \int_\delta^T F^*(t'; t_i)\, dt'.$$

This completes the proof. $\qquad\qquad\qquad\qquad\qquad\qquad\qquad\qquad\qquad\qquad\qquad\square$

Note that $\Xi \to 1$ for $T \to \infty$ since $B(T)$ is bounded. This property will later be useful to show that, given Assumption 4.15, there always exists a finite prediction horizon T large enough such that the closed-loop using the MPC controller given by Algorithm 4.30 is asymptotically stable.

Lemma 4.33 (Calculation of γ). *Suppose that Assumptions 4.1–4.3, 4.13, and Controllability Assumption 4.15 are satisfied for the nonlinear time-delay system (4.1). Then,*

$$\int_\mathcal{E}^T F^*(t'; 0) dt' \leq \gamma \int_{-\tau}^\delta F^*(t'; 0)\, dt' \tag{4.51}$$

with $\gamma = B(T) \frac{1 + \tau}{\delta}$.

Proof. Let Assumption 4.15 be satisfied. Then, the inequality

$$\int_t^T F^*(t'; 0) dt' \leq B(T - t) \left(F^*(t; 0) + \int_{t-\tau}^t F^*(t'; 0)\, dt' \right)$$

holds for all $t \in [0, T]$. This result is a direct consequence of the principle of optimality (endpieces of optimal trajectories are optimal), see (Bellman, 1957, Chapter III, §3). Furthermore,

$$\int_\delta^T F^*(t'; 0) dt' \leq \int_{t_1}^T F^*(t'; 0) dt' \qquad \text{and} \qquad B(T - t_2) \leq B(T)$$

hold for all $t_1 \in [0, \delta]$ and $t_2 \in [0, T]$ due to non-negativity of F^* and the non-decreasing property of B. By combining all of these intermediate results, we obtain

$$\int_\delta^T F^*(t'; 0) dt' \leq B(T) \min_{t \in [0,\delta]} \left(F^*(t; 0) + \int_{t-\tau}^t F^*(t'; 0) \, dt' \right).$$

Using Lemma 4.31, it follows that

$$\min_{t \in [0,\delta]} \left(F^*(t; 0) + \int_{t-\tau}^t F^*(t'; 0) \, dt' \right) \leq \frac{1+\tau}{\delta} \int_{-\tau}^\delta F^*(t'; 0) \, dt',$$

and we finally arrive at

$$\int_\delta^T F^*(t'; 0) dt' \leq \underbrace{B(T) \frac{1+\tau}{\delta}}_{=\gamma} \int_{-\tau}^\delta F^*(t'; t_i) \, dt'.$$

This completes the proof. □

It is interesting to note that the results in Lemmata 4.32 and 4.33 are more closely related to Lemmata 3.20 and 3.21 than to Lemmata 3.14 and 3.15. The techniques applied in the proofs of the latter lemmata are not suitable in the setup considered in this section. Thus, the more conservative estimates have to be used instead.

4.5.3 Asymptotic Stability and Suboptimality Estimate

Based on the results of the previous section, we can now state our main result regarding asymptotic stability of the closed-loop using the unconstrained MPC scheme for nonlinear time-delay systems as follows.

Theorem 4.34 (Stability of Unconstrained MPC for Nonlinear Time-Delay Systems). *Suppose that Assumptions 4.1–4.3, 4.13, 4.14, and Controllability Assumption 4.15 are satisfied for the nonlinear time-delay system* (4.1). *Furthermore, suppose that*

$$\alpha = 1 - (\mathcal{N} + 1)(\Xi - 1)\gamma > 0, \tag{4.52}$$

in which $\mathcal{N} = \text{ceil}\left(\frac{\tau}{\delta}\right)$ *and with* Ξ *and* γ *defined in Lemmata 4.32 and 4.33, respectively. Then, the closed-loop system resulting from the application of the model predictive controller according to Algorithm 4.30 to system* (4.1) *is asymptotically stable.*

Proof. Consider the optimal cost J_T^* at two arbitrary sampling instants t_i and t_0 with $t_i > t_0$. Adding zero and reordering terms directly yields

$$J_T^*(x_{t_i}) - J_T^*(x_{t_0}) = J_T^*(x_{t_i}) - J_T^*(x_{t_0}) + \underbrace{\sum_{j=1}^{i-1} J_T^*(x_{t_j}) - J_T^*(x_{t_j})}_{=0}$$

$$= \sum_{j=0}^{i-1} J_T^*(x_{t_{j+1}}) - J_T^*(x_{t_j}). \tag{4.53}$$

Using Lemma 4.32, we obtain

$$J_T^*(x_{t_{j+1}}) - J_T^*(x_{t_j}) = J_T^*(x_{t_j+\delta}) - J_T^*(x_{t_j})$$
$$\overset{(4.49)}{\leq} (\Xi - 1) \int_{t_j+\delta}^{t_j+T} F^*(t'; t_j) dt' - \int_{t_j}^{t_j+\delta} F^*(t'; t_j) dt'.$$

Moreover, it follows from Lemma 4.33 that

$$J_T^*(x_{t_{j+1}}) - J_T^*(x_{t_j}) \overset{(4.51)}{\leq} ((\Xi - 1)\gamma - 1) \int_{t_j}^{t_j+\delta} F^*(t'; t_j) dt' + (\Xi - 1)\gamma \int_{t_j-\tau}^{t_j} F^*(t'; t_j) dt'.$$
$$(4.54)$$

Since we consider no model plant mismatch, the predicted trajectories and the actual trajectories of the closed-loop system coincide until the next sampling instant. Thus, $F^*(t'; t_j) = F(x(t'), u_{\mathrm{MPC}}(t'))$ for $t' \in [t_j, t_j + \delta]$. Careful inspection of the sum in (4.53) in combination with (4.54) then yields

$$J_T^*(x_{t_i}) - J_T^*(x_{t_0})$$
$$\leq ((\Xi - 1)\gamma - 1) \int_{t_0}^{t_i} F(x(t'), u_{\mathrm{MPC}}(t')) \, dt'$$
$$+ \mathcal{N}(\Xi - 1)\gamma \int_{t_0}^{t_i} F(x(t'), u_{\mathrm{MPC}}(t')) \, dt' + \mathcal{N}(\Xi - 1)\gamma \int_{t_0-\tau}^{t_0} F^*(t'; t_0) \, dt'$$
$$= \underbrace{((\mathcal{N} + 1)(\Xi - 1)\gamma - 1)}_{=-\alpha} \int_{t_0}^{t_i} F(x(t'), u_{\mathrm{MPC}}(t')) dt' + \mathcal{N}(\Xi - 1)\gamma \int_{t_0-\tau}^{t_0} F^*(t'; t_0) \, dt'.$$

Since $J_T^*(x_{t_i}) > 0$ and $J_T^*(x_{t_0})$ are finite, it follows for any arbitrary $t_i > t_0$

$$\int_{t_0}^{t_i} F(x(t'), u_{\mathrm{MPC}}(t')) \, dt' \leq \frac{1}{\alpha} J_T^*(x_{t_0}) + \mathcal{N}(\Xi - 1)\frac{\gamma}{\alpha} \int_{t_0-\tau}^{t_0} F^*(t'; t_0) \, dt' < \infty. \qquad (4.55)$$

Asymptotic stability follows directly from standard arguments in optimal control and Barbalat's Lemma (Barbalat, 1959; Khalil, 2002). This completes the proof. □

Note that $\Xi \to 1$ for $T \to \infty$, which directly implies $\alpha \to 1$ for $T \to \infty$ as in the delay-free case, see Proposition 3.17. Hence, there always exists a finite prediction horizon T chosen suitably large such that the closed-loop using the MPC controller is asymptotically stable.

Regarding suboptimality estimates, we can make the following comments. Equation (4.55) and $J_T^\star(\varphi) \leq J_\infty^\star(\varphi)$ yield for initial condition φ

$$J_\infty^\star(\varphi) \leq J_\infty^{\mathrm{MPC}}(\varphi) = \int_0^\infty F(x(t'), u_{\mathrm{MPC}}(t'))\, dt'$$

$$\leq \frac{1}{\alpha} J_T^\star(\varphi) + \mathcal{N}(\Xi - 1)\frac{\gamma}{\alpha}\int_{-\tau}^0 F(\varphi(t'), 0)\, dt' \tag{4.56a}$$

$$\leq \frac{1}{\alpha} J_\infty^\star(\varphi) + \mathcal{N}(\Xi - 1)\frac{\gamma}{\alpha}\int_{-\tau}^0 F(\varphi(t'), 0)\, dt'. \tag{4.56b}$$

Thus, an upper bound on the infinite horizon performance of the MPC controller can be given in terms of the finite horizon optimal cost at initial time $J_T^\star(\varphi)$ and on an additional term depending on the initial state φ. This second term is not necessary in the case of delay-free systems as can be directly seen for $\tau = 0$. Furthermore, Equation (4.56) shows that $J_\infty^{\mathrm{MPC}}(\varphi) \to J_\infty^\star(\varphi)$ for $T \to \infty$, i.e., infinite horizon optimal performance is recovered for large enough prediction horizon and the influence of the second term in (4.56b), which depends on the initial condition φ, vanishes.

Moreover, we obtain $\mathcal{N} = 0$ and $\alpha = 1 - \gamma(\Xi - 1)$ for $\tau = 0$. This directly recovers the stability condition of Theorem 3.16 for delay-free continuous-time systems, albeit with the "worse" estimates $\Xi^\dagger$ and $\gamma^\dagger$ from Lemmata 3.20 and 3.21. Furthermore, Equation (4.55) becomes

$$\int_{t_0}^\infty F(x(t'), u_{\mathrm{MPC}}(t'))\, dt' \leq \frac{1}{\alpha} J_T^\star(x_{t_0}) \leq \frac{1}{\alpha} J_\infty^\star(x_{t_0}),$$

which recovers the suboptimality estimate of the infinite horizon performance of the MPC controller without additional terms depending on the initial condition, see also Theorem 3.16.

Note that $\alpha \to -\infty$ for $\delta \to 0$, i.e., asymptotic stability of the closed-loop cannot be guaranteed for arbitrarily small sampling times, which is to a certain extent counterintuitive. However, it was shown in Grüne et al. (2010b,c) and Section 3.2.3 that with an additional condition, the so-called *growth condition*, this effect can be avoided for discrete-time and continuous-time delay-free systems. The estimate obtained in Section 3.2.3 satisfies $\Xi \to 1$ for $\delta \to 0$, which allows to cancel the effect of $\gamma \to \infty$ for $\delta \to 0$ when using the growth condition. Unfortunately, this is not possible for the results presented in this section. It is simple to see in Lemma 4.32, that $\Xi \not\to 1$ for $\delta \to 0$, independently of $B(T)$. Hence, the growth condition is not applicable to avoid the poor estimates for small sampling times.

It is also interesting to note that stability is not proven by a decrease of the optimal cost function from one sampling instant to the next one, but only by a decrease in the long run. The optimal cost can indeed increase along trajectories of the closed-loop due to the effect of the delayed states.

4.5.4 Summary

In this section, we considered model predictive control for nonlinear time-delay systems using neither terminal constraints nor control Lyapunov-Krasovskii functionals as terminal weighting terms. First, we proposed an extended asymptotic controllability assumption, which was necessary because the stage cost only penalizes the instantaneous state instead of the full delayed state. Based on this assumption, we provided conditions on the length of the prediction horizon to guarantee nominal asymptotic stability of the closed-loop. In contrast to most other results on stability of MPC (regardless whether systems with or without time-delay are considered), the optimal cost is not used as Lyapunov function or Lyapunov-Krasovskii functional. The optimal cost can indeed increase along trajectories of the closed-loop due to the effect of the delayed states, but has to decrease in the long run. However, poor estimates for small sampling times are obtained and a growth condition similar to Section 3.2.3 cannot be used for improving these estimates due to the techniques required by the presence of time-delays.

4.6 Numerical Examples

In this section, we compare the MPC schemes for nonlinear time-delay systems presented in this chapter by using two numerical examples. In Section 4.6.1, we show simulation results for the simple scalar nonlinear time-delay system (4.11), which was already used in several examples in Section 4.3. In Section 4.6.2, we investigate a more realistic example by considering the stabilization of an unstable equilibrium of a continuous stirred tank reactor with recycle stream.

4.6.1 Simple Scalar Example

In this section, we consider the simple scalar nonlinear time-delay system (4.11), i.e., $\dot{x}(t) = x(t - \tau)^4 + u(t)$ with time-delay $\tau = 1$. This system was introduced in Example 4.6 and also considered in Examples 4.9, 4.12, 4.18, and 4.22. We show simulation results for six different MPC setups. First, we apply MPC with terminal cost and terminal constraints, as described in Section 4.2, with the design parameters calculated by the four procedures introduced in Section 4.3. The terminal cost and terminal region for each procedure are also summarized in Table 4.2 in Section 4.3.5. Second, we consider the two MPC schemes without terminal constraints which were derived in Sections 4.4 and 4.5, respectively. The terminal cost for the unconstrained MPC scheme in Section 4.4 is chosen based on the design by Lyapunov-Razumikhin arguments with additional exponential weighting, see Section 4.3.4 and further explanations below.

For our simulations, we choose the initial condition $\varphi(\theta) = 1.5 + 3\theta$ for $\theta \in [-\tau, 0]$ and input constraints $\mathbb{U} = [-4, 4]$. Moreover, we choose the prediction horizon $T = 2$ and the sampling time $\delta = 0.1$. For the design with Lyapunov-Razumikhin arguments without additional weighting in Section 4.3.3, we require $\delta > \tau$, see Assumption 4.9, hence, choosing $\delta = 1.1$ for this scheme and, consequently, the parameter $\beta = 11$ in the terminal cost $E(x_t)$, see Table 4.2.

The simulation results are shown in Figure 4.8. For this comparatively simple system, only negligible differences are visible in the evolution of the states and all schemes asymptotically stabilize the system and satisfy the input constraints. The most significant difference compared to the other schemes can be seen for the MPC scheme based on Section 4.3.3, which is mainly due to the use of a different sampling time $\delta = 1.1$ in contrast to $\delta = 0.1$ in all other schemes. As could be expected, unconstrained MPC without terminal constraints needs the least computation time with about 30% less computation time compared to the MPC schemes with terminal constraints. Out of the MPC schemes with terminal constraints, the design with Lyapunov-Razumikhin and exponential weighting required about 14% less computation time than the schemes with terminal cost functionals based on Lyapunov-Krasovskii arguments. While this indicates advantages of using Razumikhin-arguments if possible, a more thorough investigation of numerical and implementation aspects for different examples would be necessary for stronger statements. However, this is beyond the scope of this thesis.

The possible robustness problems of the MPC scheme based on Section 4.3.3 are briefly illustrated in Figure 4.9. We compare the MPC scheme based on Section 4.3.1 with sampling time $\delta = 0.1$ and the MPC scheme based on Section 4.3.3 with sampling time $\delta = 1.1$ for the initial condition $\varphi(\theta) = 1.3$ for $\theta \in [-\tau, 0]$. We consider a model plant mismatch, the nominal model used for the predictions in both MPC schemes is given by (4.11), i.e., $\dot{x}(t) = x(t-\tau)^4 + u(t)$, and the actual system is described by $\dot{x}(t) = x(t-\tau)^4 + 0.2x(t)^2 + u(t)$. For the sampling time $\delta = 0.1$, the inherent robustness of the MPC suffices to asymptotically stabilize the system. However, for $\delta = 1.1$ the closed-loop does not converge to the origin.

4.6.2 Continuous Stirred Tank Reactor with Recycle Stream

In this section, we consider the model of a continuous stirred tank reactor with recycle stream and the stabilization of an unstable steady state subject to input constraints. Model and parameters are taken from the example of Findeisen and Allgöwer (2000) and the model is extended with a recycle stream, see Figure 4.10. The equations of the reactor following from the mass and energy balance are given by

$$\dot{c}(t) = a_1 \left(c_{in}(t) - c(t)\right) - 2\,K(\mathfrak{T}(t))\,c(t)^2 \tag{4.57a}$$

$$\dot{\mathfrak{T}}(t) = a_1 \left(\mathfrak{T}_{in}(t) - \mathfrak{T}(t)\right) + a_2 \left(\mathfrak{T}_k(t) - \mathfrak{T}(t)\right) + a_3\,K(\mathfrak{T}(t))\,c(t)^2 \tag{4.57b}$$

in which

$$c_{in}(t) = (1-\nu)\,c_f + \nu\,c(t-\tau) \quad \text{and} \quad \mathfrak{T}_{in}(t) = (1-\nu)\,\mathfrak{T}_f + \nu\,\mathfrak{T}(t-\tau)\,.$$

The temperature and concentration of the reactant inside the reactor are denoted by $\mathfrak{T}$ and c, respectively. $\mathfrak{T}_f$ and c_f are the temperature and concentration of the inflow and both are assumed to be constant. The manipulated input is the heating jacket temperature $\mathfrak{T}_k$. The coefficient $\nu \in [0,1]$ is the recirculation coefficient and τ the recycle time. The rate of reaction $K(\mathfrak{T})$ is given by the Arrhenius law $K(\mathfrak{T}) = k_0\,e^{-\frac{a_4}{\mathfrak{T}}}$. The other model parameters are

$$a_1 = \frac{q}{V}\,, \quad a_2 = \frac{k_w F_k}{\rho c_p V}\,, \quad a_3 = \frac{-\Delta h_r}{\rho c_p}\,, \quad a_4 = \frac{E_A}{R}\,.$$

For our simulations, we use the following numerical values of the parameters:

$$\nu = 0.5\,, \ \tau = 20\,\mathrm{s}\,, \ q = 0.1\,\frac{1}{\mathrm{min}}\,, \ V = 1000\,\mathrm{cm}^3 = 1\,\mathrm{l}\,,$$

$$k_w = 0.1\,\frac{\mathrm{cal}}{\mathrm{cm}^2\,\mathrm{min\,K}}\,, \ F_k = 250\,\mathrm{cm}^2\,, \ \rho c_p = 0.659\,\frac{\mathrm{cal}}{\mathrm{cm}^3\,\mathrm{K}}\,,$$

$$\Delta h_r = -20000\,\frac{\mathrm{cal}}{\mathrm{mol}}\,, \ E_A = -\Delta h_r\,, \ R = 1.9864\,\frac{\mathrm{cal}}{\mathrm{mol\,K}}\,, \ k_0 = 33 \cdot 10^9\,\frac{1}{\mathrm{mol\,min}}\,.$$

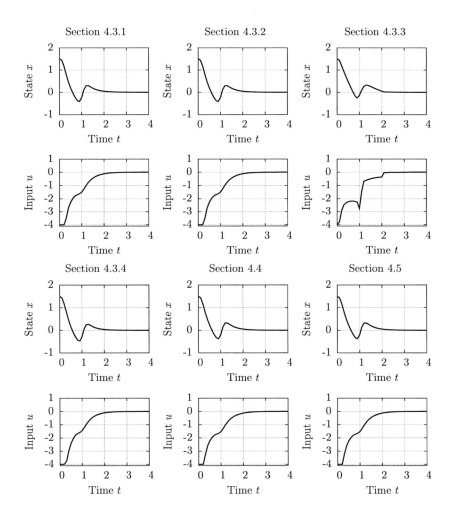

Figure 4.8: Simulation results for the simple scalar nonlinear time-delay system (4.11) discussed in Section 4.6.1. From top left to bottom right: MPC with terminal constraints based on four different design schemes (see Sections 4.3.1–4.3.4), unconstrained MPC with terminal cost (Section 4.4) and unconstrained MPC without terminal cost (Section 4.5).

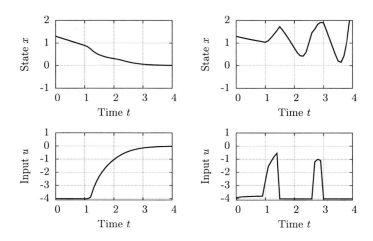

Figure 4.9: Simulation results for the simple scalar nonlinear time-delay system (4.11) with uncertainty. Left: MPC scheme based on Section 4.3.1 with $\delta = 0.1$. Right: MPC scheme based on Section 4.3.3 with $\delta = 1.1$.

The goal is to stabilize the unstable steady state $\mathfrak{T}_s = 345$ K, $c_s = 4.24$ mol/l for constant inflow parameters $\mathfrak{T}_f = 290$ K, $c_f = 6.67$ mol/l, and heating jacket temperature $\mathfrak{T}_{k,s} = 389$ K. Hence, we define

$$x_1 = c(t) - c_s\,, \quad x_2 = \mathfrak{T}(t) - \mathfrak{T}_s, \quad \text{and} \quad u = \mathfrak{T}_k - \mathfrak{T}_{k,s}\,.$$

The input $\mathfrak{T}_k$ is constrained between 349 K and 429 K, i.e., $|u| = |\mathfrak{T}_k - \mathfrak{T}_{k,s}| \leq 40$ K and $\mathbb{U} = [-40\ \text{K}, 40\ \text{K}]$.

In order to apply the results of Sections 4.3.1–4.3.4 the Jacobi linearization about the steady state is calculated and the resulting LMIs are solved in MATLAB using YALMIP (Löfberg, 2004). The weighting matrices are chosen as $Q = 100\,I$ and $R = I$. The resulting local control law is $u(t) = \mathfrak{T}_k(x_t) - \mathfrak{T}_{k,s}$ with

$$\mathfrak{T}_k(x_t) - \mathfrak{T}_{k,s} = Kx(t) = [-49.18\frac{\text{K1}}{\text{mol}}, \ -26.41] \begin{bmatrix} c(t) - c_s \\ \mathfrak{T}(t) - \mathfrak{T}_s \end{bmatrix}.$$

The solutions of the LMIs then give conditions on a sufficiently small γ in (4.8), see, e.g., Equations (4.21) and (A.22). For γ determined this way, it remains to calculate δ_γ such that Property (4.8) holds for all $\|x_t\|_\tau \leq \delta_\gamma$ and $|u(t)| < \delta_\gamma$. Once δ_γ is obtained, the terminal region Ω can be calculated in order to ensure $\|x_t\|_\tau \leq \delta_\gamma$, $|Kx(t))| < \delta_\gamma$, and $Kx(t) \in \mathbb{U}$ for all $x_t \in \Omega$. One possible approach for the calculation of δ_γ and Ω is described in the following. Φ only consists of higher order terms and does not contain any delayed terms for the model of the CSTR (4.57). Hence, we can define a function $\tilde{\Phi}(x)$ such that $\Phi(x_t, Kx(t)) = \tilde{\Phi}(x(t))$. Due to the residual of the Taylor series expansion, we know that $\tilde{\Phi}(x) = \frac{1}{2!}(\xi x)^T \mathcal{H}\tilde{\Phi}(\xi x)(\xi x)$ for some $\xi \in [0,1]$ (Bronstein et al., 2000). Here,

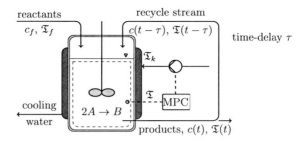

reactants
c_f, $\mathfrak{T}_f$

recycle stream
$c(t-\tau)$, $\mathfrak{T}(t-\tau)$

time-delay τ

$\mathfrak{T}_k$

cooling
water

$2A \to B$

$\mathfrak{T}$
MPC

products, $c(t)$, $\mathfrak{T}(t)$

Figure 4.10: Continuous stirred tank reactor with recycle stream (source: Hasenauer (2012)).

$\mathcal{H}\tilde{\Phi}(\xi x)$ denotes the Hessian matrix of $\tilde{\Phi}$ with respect to x evaluated at ξx. By using an upper bound $\widehat{\mathcal{H}}$ on the Hessian matrix $\widehat{\mathcal{H}}$ in some region $\bar{\Omega}$ around the origin, we obtain $|\tilde{\Phi}(x)| \leq \frac{1}{2}\widehat{\mathcal{H}}|x|^2$ in this origin. Now by choosing α and δ_γ small enough, it is possible to guarantee that $|x(t)| \leq \frac{2\gamma}{\widehat{\mathcal{H}}}$ for all states x_t inside the terminal region $\Omega \in \bar{\Omega}$. Thus, $|\tilde{\Phi}(x)| \leq \gamma|x|$ and consequently Property (4.8) is satisfied for all states inside the terminal region for the desired γ. If a smaller upper bound $\widehat{\mathcal{H}}$ can be calculated in Ω, we can repeat this procedure starting from a smaller region $\bar{\Omega}$. A smaller upper bound $\widehat{\mathcal{H}}$ calculated for this newly chosen $\bar{\Omega}$ then yields a larger terminal set Ω. This can be repeated iteratively until $\bar{\Omega} = \Omega$. Using this approach, the terminal region can be calculated for each of the procedures of Sections 4.3.1–4.3.4.

The simulation results for a prediction horizon of $T = 35$ minutes are shown in Figure 4.11. As can be expected the model predictive controller stabilizes the unstable steady state $\mathfrak{T}_s$, c_s while satisfying the input constraints. In this example, the terminal region is relatively small for the procedures which explains the similar behavior of the different controllers and shows the still existing conservatism in the proposed schemes.

One reason for this is that results based on the Jacobi linearization also lead to conservative results in the delay-free case. This is due to the fact that often only conservative bounds on the nonlinearity have to be used such as (4.8), as well as the restriction to quadratic Lyapunov functions which might not be appropriate for nonlinear systems. Furthermore, we have only used the simplest quadratic Lyapunov-Krasovskii functional with constant matrices P and S in order to calculate the local control law as a first step. However, it is possible to generalize the principle ideas of using a local control law and either an additional Razumikhin condition or the terminal region defined by the intersection of a sublevel set and a norm bounded region in $\mathcal{C}_\tau$. For instance, one future step can be the consideration of more complicated functionals, e.g., calculated by means of sum-of-squares techniques as in Papachristodoulou (2005); Papachristodoulou et al. (2005).

In Figure 4.12, we show simulation results for MPC schemes without terminal constraints. We choose the terminal cost again based on the design by Lyapunov-Razumikhin arguments with additional exponential weighting. In this example, both unconstrained MPC schemes stabilize the system and exhibit similar performance of the closed-loop compared to the results with terminal constraints.

4.7 Summary

In this chapter, we proposed different MPC schemes for nonlinear time-delay systems with guaranteed stability of the closed-loop.

First, we extended the classical well-known MPC scheme using a terminal cost functional and terminal constraints to the problem setup with time-delays. While the stability conditions are very similar to the delay-free case, the calculation of suitable design parameters based on the Jacobi linearization about the origin was shown to be more difficult due to the infinite-dimensional nature of the state space. We proposed four different schemes to overcome these difficulties, each of which contains the delay-free case as special case, and discussed the properties of each of these schemes. Second, we presented two MPC schemes without terminal constraints. Last, we compared the different MPC schemes for two numerical examples. For these examples, similar performance was observed for all MPC schemes. The general analysis carried out in this chapter yields rather conservative stability conditions and, in particular, the computation time seems to be still prohibitive for practical applications. However, we see our results as a useful contribution towards a better fundamental understanding of MPC for time-delay systems and as a theoretical basis for future developments in this area.

For open questions and possible directions of future research, we refer to Section 5.2 at the end of this thesis.

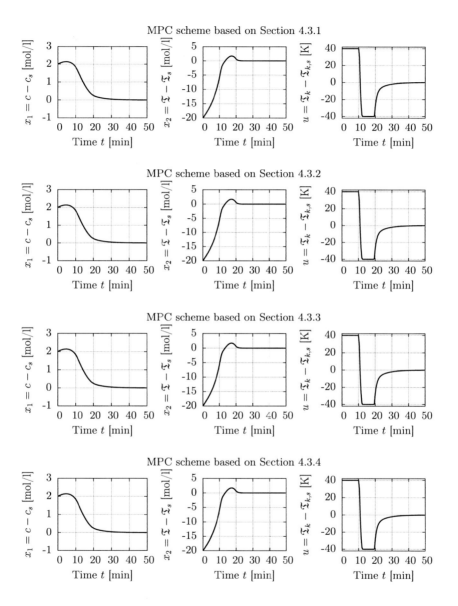

Figure 4.11: Simulation results for the CSTR (4.57) using MPC with terminal constraints.

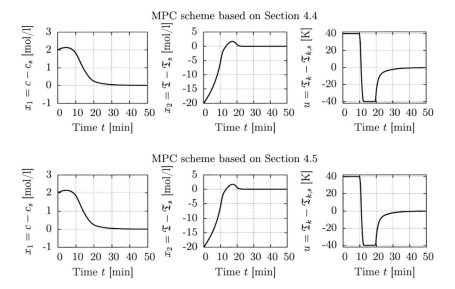

Figure 4.12: Simulation results for the CSTR (4.57) using MPC without terminal constraints.

Chapter 5

Conclusions

In this chapter, we summarize the main results of the thesis and indicate possible directions for future research.

5.1 Summary and Discussion

In this thesis, we derived novel stability conditions for model predictive control for nonlinear continuous-time systems with and without time-delays.

In Chapter 3, we considered different MPC schemes based on a controllability assumption for finite-dimensional nonlinear continuous-time systems. This controllability assumption is less restrictive than the usual assumption in MPC of knowing a local control Lyapunov function. Hence, these MPC schemes are attractive alternative control methods especially for systems for which no local control Lyapunov function is known and/or without stabilizable Jacobi linearization such as the nonholonomic or the cubic integrator. We derived explicit conditions on the prediction horizon such that closed-loop stability is guaranteed for the different MPC schemes, in particular for unconstrained MPC without terminal cost. This simplest possible MPC setup is important due to its widespread application in industry, and also has the advantage of giving guaranteed performance bounds for the closed-loop. Starting from this MPC scheme, we have shown that additional weighting terms can be used in order to guarantee stability for shorter prediction horizons. Similarly, additional weighting terms allow to guarantee stability if only a local controllability assumption is satisfied in combination with appropriate terminal cost terms. It is interesting to note that these MPC schemes show connections between unconstrained MPC and the classical MPC schemes with terminal cost and terminal constraint, two classes of MPC schemes which have been considered separately in the literature so far. More specifically, both classes of MPC schemes can be recovered as a limit case of the unifying framework presented in Section 3.5, which uses an integral terminal cost and extended terminal constraints. Furthermore, this setup allows more flexibility and can be useful in some cases for which both of the previous MPC schemes are not applicable.

In Chapter 4, we examined MPC for nonlinear time-delay systems. We have derived rigorous stability conditions for several MPC setups with and without terminal cost terms and/or terminal constraints for this class of systems. While the stability conditions are similar to those for delay-free systems, the actual calculation of stabilizing design parameters requires additional arguments. For the examples studied in this chapter, namely an academic example and a continuous stirred tank reactor with recycle stream, similar performance was observed for all MPC schemes. Our analysis yields rather general results with relatively few assumptions on the system, which can result in conservative stability conditions. In particular, a rather small terminal region might be obtained which requires a large computation time and might be prohibitive for direct application in practical examples.

In all MPC schemes considered in this thesis, the optimal cost is used as a Lyapunov function of the closed-loop – with the notable exception of the results on unconstrained MPC for time-delay systems in Section 4.5.

5.2 Outlook

Only nominal stability of a known steady state has been addressed in this thesis. Therefore, other properties of the closed-loop and the applicability in different areas are natural future research directions for the MPC schemes based on a controllability assumption discussed in Chapter 3. For some special cases, first results are available such as unconstrained MPC for networked control systems (Grüne et al., 2012; Reble et al., 2012c), for distributed systems (Grüne and Worthmann, 2012), for economic MPC (Grüne, 2011, 2012), and for path following problems (Faulwasser, 2012). Other topics have not been investigated so far, such as robustness properties, the design of robust MPC schemes, and extensions towards stochastic MPC.

Another particularly challenging, yet crucial, problem is the rigorous analysis of efficient suboptimal MPC schemes. For the classical MPC schemes using a local CLF and terminal constraints, it is well-known that "feasibility implies stability" to a certain extent (Scokaert et al., 1999). A feasible candidate solution to the open-loop optimal control problem is easily constructed by appending control values based on the local control law to the solution at the previous sampling time. Stability is then guaranteed for all feasible, not necessarily optimal, solutions which yield a smaller cost than this candidate solution. Close inspection reveals that similar arguments do not work for the stability analysis in unconstrained MPC. For instance, the calculation of the suboptimality estimate in Theorem 3.16 relies on the fact that the optimal cost is smaller than the cost associated to each of infinitely many feasible candidate solutions provided by the controllability assumption. Hence, it is more difficult to find a feasible solution which guarantees a decrease of the optimal cost and thereby stability of the closed-loop. For a more detailed analysis of suboptimal MPC, see, e.g., Diehl et al. (2004); Graichen and Kugi (2010).

Furthermore, it would be interesting to investigate computational aspects more closely. For standard MPC problems, there now exists a huge variety of specifically tailored numerical methods, e.g., (Diehl et al., 2004; Graichen, 2012; Ohtsuka, 2004; Sideris and Bobrow, 2005) to name only a few. However, several different non-standard finite horizon optimal control problems have been proposed for use in MPC schemes in Chapter 3, e.g., the setup using a generalized integral cost and a generalized terminal constraint in Section 3.5. The formulation and performance of numerical methods for these non-standard problems is an open question so far.

A problem related to MPC for nonlinear time-delay systems is the control of other classes of infinite-dimensional systems, see Section 2.1.2 for several references on MPC for PDEs. In most of the literature, a rigorous stability analysis is only carried out for linear PDEs or MPC with a global control Lyapunov functional as terminal cost. However, terminal inequality constraints have rarely been used for PDEs. The design of these constraints based on the Jacobi linearization about the steady state would be subject to similar difficulties as for nonlinear time-delay systems due to the infinite-dimensional nature of the system. Hence, the procedures proposed in Section 4.3 would be a good starting point in order to overcome these difficulties.

Furthermore, it would be desirable to have better stability conditions regarding unconstrained MPC for nonlinear time-delay systems, in particular for small sampling times. The results presented in Section 4.5 cannot be directly improved by a growth condition as in Section 3.2 for delay-free systems. Hence, different analysis techniques are necessary. In this respect, note that other MPC setups for time-delay systems are also possible which do not belong to the framework presented in Chapter 4. For instance, an MPC setup in which the delayed states are taken into account in the definition of the stage cost is presented in Brunner (2010).

Another very interesting field of future research is robust MPC for nonlinear time-delay systems. Due to the high computational demand even for nominal MPC, min-max MPC (Raimondo et al., 2009) is not an appropriate choice for a robust MPC design. Tube-based MPC (Langson et al., 2004; Mayne et al., 2005) appears to be a more suitable starting point for research in this area.

Finally, it would be worthwhile to investigate output feedback for nonlinear time-delay systems. While there are several results available concerning the state estimation and observer design for nonlinear time-delay systems (Boutayeb, 2001; Germani et al., 1998, 1999, 2001; Koshkouei and Burnham, 2009; Raff and Allgöwer, 2006; Zemouche et al., 2007), there exist only very few results on separation principles with one exception for a particular setup given by Germani et al. (2010). A good starting point would be to consider the results on nonlinear separation principles for finite-dimensional systems, see, e.g., the results of Arcak (2002); Atassi and Khalil (1999, 2000, 2001); Teel and Praly (1994), and the results on output feedback using MPC, see, e.g., the results of Findeisen (2004); Findeisen et al. (2003); Roset et al. (2006). In order to overcome the problems raised by the infinite-dimensional state space, the techniques used in Section 4.3 seem to be promising.

Appendix A

Technical Proofs

A.1 Proof of Theorem 4.5

The proof of Theorem 4.5 follows the lines of Chen (1997), however, the infinite-dimensional nature of the state space requires proof of additional properties of the optimal value functional, see Lemma A.3, which are directly implied by continuity and positive definiteness in the finite-dimensional case.

First, we address feasibility of the open-loop optimal control problem in Lemma A.1. To establish asymptotic stability, it will be then shown in Lemma A.2 that the optimal cost $J_T^*(x_{t_i})$ of Problem 4.3 is continuous in the state x_t and is locally lower bounded as shown in Lemma A.3. Continuity of the optimal cost is required for the proof of asymptotic stability as opposed to only convergence. Furthermore, the optimal value functional is non-increasing along trajectories of the closed-loop as proven in Lemma A.5. In the last step, asymptotic stability is shown using these intermediate results.

Lemma A.1. *Suppose that Assumptions 4.1–4.6 are satisfied. Then, the open-loop finite horizon optimal control problem 4.3 admits a feasible solution for all times $t \in \mathbb{R}_{>0}$, if it is initially feasible at time $t_i = 0$.*

Proof. Suppose that at time t_i, a feasible solution of Problem 4.3 exists and denote this solution by $u(t'; x_{t_i}, t_i)$, $t' \in [t_i, t_i + T]$. At time $t^* \in]t_i, t_i + \delta]$, a feasible, but not necessarily optimal, control input $\hat{u}$ can be constructed by appending control values based on the local controller $k(x_t)$ to the solution at the previous sampling time t_i. Formally, $\hat{u}$ is defined by

$$\hat{u}(t'; x_{t_i}, t_i) = \begin{cases} u(t'; x_{t_i}, t_i), & \text{for } t' \in [t^*, t_i + T[\\ k(\bar{x}_{t'}), & \text{for } t' \in [t_i + T, t^* + T] \end{cases} \tag{A.1}$$

in which $\bar{x}_{t'}$ is defined by the prediction of the closed-loop

$$\begin{aligned} \dot{\bar{x}}(t') &= f(\bar{x}(t'), \bar{x}(t' - \tau), u(t'; x_{t_i}, t_i)) && \text{for } t' \in [t^*, t_i + T[, \\ \dot{\bar{x}}(t') &= f(\bar{x}(t'), \bar{x}(t' - \tau), k(\bar{x}_{t'})) && \text{for } t' \in [t_i + T, t^* + T], \\ \bar{x}_{t_i} &= x_{t_i}. \end{aligned}$$

Hence, $\hat{u}$ consists of two parts: The first part is the feasible control calculated at time t_i, which steers the system from x_{t^*} to $x_{t_i + T} \in \Omega$ inside the terminal region and the second part uses the local controller $k(x_t)$, which keeps the system trajectory in Ω for $t_i + T \leq t' \leq t^* + T$ while respecting the input constraints. Hence, the feasibility of Problem 4.3 at time t_i implies feasibility at time t^* and in particular also at the next sampling instant $t_{i+1} = t_i + \delta$. By induction, Problem 4.3 is feasible for every $t > 0$ if it is feasible at initial time $t = 0$. □

Lemma A.2. *Suppose that Assumptions 4.1–4.6 are satisfied. Then, the optimal cost $J_T^*(x_t)$ of the open-loop finite horizon optimal control problem 4.3 is continuous in x_t at $x_t = 0$.*

Proof. Let $\varphi \in \mathcal{C}_\tau$ belong to some neighborhood $\mathfrak{C} \subseteq \mathcal{C}_\tau$ of the origin and $\varphi \neq 0$. Choose $\bar{u} = 0$ as a candidate solution to the finite horizon optimal control problem 4.3. Now, consider the following system for $t' \in [t, t+T]$

$$\dot{\bar{x}}(t') = f(\bar{x}(t'), \bar{x}(t'-\tau), 0), \quad x_t = \varphi.$$

If the neighborhood $\mathfrak{C}$ is chosen sufficiently small, a unique solution $\bar{x}(t)$ exists on the interval $[t, t+T]$ because f is a continuously differentiable function of its arguments. This solution depends continuously on the initial condition φ, see, e.g., (Hale and Lunel, 1993, Theorem 2.2) and (Kolmanovskii and Myshkis, 1999). Furthermore, we can choose $\mathfrak{C}$ sufficiently small such that $\bar{x}_{t+T}$ is inside the terminal region because 0 is in the interior of Ω. Now, let the associated cost functional be denoted by

$$\bar{J}_T^*(\varphi) = \int_t^{t+T} F(\bar{x}(t'), 0)dt' + E(\bar{x}_{t+T}). \tag{A.2}$$

Since F and E are continuous and $x(t)$ depends continuously on φ in a neighborhood of $x_t = 0$, $\bar{J}_T^*(\varphi)$ is continuous with respect to φ in this neighborhood. Thus, for any $\varepsilon > 0$, there exists δ_ε such that $\|\varphi\|_\tau < \delta_\varepsilon$ implies $\left|\bar{J}_T^*(\varphi)\right| < \varepsilon$. On the other hand, the optimal input $u_T^*(t'; x_{t_i}, t_i)$ will yield no larger cost than the selected candidate solution $\bar{u} = 0$ and $J_T^* \geq 0$. Hence, for $\|\varphi\|_\tau < \delta_\varepsilon$, we have

$$|J_T^*(\varphi)| \leq \left|\bar{J}_T^*(\varphi)\right| < \varepsilon.$$

Thus, $J_T^*(\varphi)$ is continuous at the origin $x_t = 0$. $\qquad\square$

Lemma A.3. *The optimal value functional $J_T^*(x_t)$ of the open-loop finite horizon optimal control problem 4.3 satisfies*

$$J_T^*(x_t) \geq \underline{\alpha}_J(|x(t)|) \tag{A.3}$$

in a neighborhood of the origin, in which $\underline{\alpha}_J : \mathbb{R}_{\geq 0} \to \mathbb{R}_{\geq 0}$ is a class $\mathcal{K}_\infty$ function.

Proof. For $\alpha \in \mathbb{R}_{>0}$, consider two regions around the origin defined by

$$\Omega_1 = \{x_t : \|x_t\|_\tau \leq \alpha\} \quad \text{and} \quad \Omega_2 = \{x_t : \|x_t\|_\tau \leq 2\alpha\}.$$

Since f is continuous and $\mathbb{U}$ is compact, there exists a positive constant $M \in \mathbb{R}_{>0}$ such that f is bounded by $|f| < M$ for all x_t in Ω_1 and Ω_2. Now let $x_t \in \Omega_1 \subset \Omega_2$. Then, clearly $|x(t)| \leq \alpha$ by definition of Ω_1 and

$$\frac{|x(t)|}{2} \leq |x(t')| \leq \frac{3|x(t)|}{2} \leq 2\alpha, \quad \forall t' \in [t, t+T_M]$$

holds for $T_M = |x(t)|/2M$. On account of (4.3) and since $E \geq 0$, it directly follows that

$$J_T^*(x_t) \geq \int_t^{\min\{T_M, T\}} \underline{\alpha}_F(|x(t')|) \, dt' \geq \underline{\alpha}_F\left(\frac{|x(t)|}{2}\right) \cdot \min\left\{\frac{|x(t)|}{2M}, T\right\}.$$

Hence, (A.3) is satisfied for

$$\underline{\alpha}_J = \underline{\alpha}_F\left(\frac{|x(t)|}{2}\right) \cdot \min\left\{\frac{|x(t)|}{2M}, T\right\},$$

which completes the proof. $\qquad\square$

Remark A.4. *Lemma A.3 is the main difference of this proof of Theorem 4.5 compared to the results in Chen (1997) for finite-dimensional systems, for which (A.3) is directly satisfied.*

Lemma A.5. *Suppose that Assumptions 4.1–4.6 are satisfied. For any sampling instant $t_i = i\,\delta$ and any $t^* \in [t_i, t_i + \mathcal{E}]$, the optimal value functional satisfies*

$$J_T^*(x_{t^*}) \leq J_T^*(x_{t_i}) - \int_{t_i}^{t^*} F(x(t'), u(t'))\, dt'. \tag{A.4}$$

Proof. Feasibility of the optimization problem is guaranteed by Lemma A.1. Let $x^*(t') = x_T^*(t'; x_{t_i}, t_i)$ denote the state resulting from application of the optimal input $u_T^*(t') = u_T^*(t'; x_{t_i}, t_i)$ starting from x_{t_i} at time t_i. The value of the objective functional at time t_i is

$$J_T^*(x_{t_i}) = \int_{t_i}^{t^*} F(x^*(t'), u_T^*(t'))\, dt' + \int_{t^*}^{t_i+T} F(x^*(t'), u_T^*(t'))\, dt' + E(x_{t_i+T}^*). \tag{A.5}$$

Let $\widehat{x}(t')$ denote the state resulting from application of the feasible, but not necessarily optimal, input (A.1) starting at $x_{t^*}^*$ at time t^*. The value of the objective functional at time t^* for this suboptimal input reads

$$J_T(x_{t^*}, \widehat{u}) = \int_{t^*}^{t^*+T} F(\widehat{x}(t'), \widehat{u}(t'))\, dt' + E(\widehat{x}_{t^*+T})$$
$$= \int_{t^*}^{t_i+T} F(x^*(t'), u_T^*(t'))\, dt' + \int_{t_i+T}^{t^*+T} F(\widehat{x}(t'), k(\widehat{x}(t')))\, dt' + E(\widehat{x}_{t^*+T}). \tag{A.6}$$

Combining (A.5), (A.6) and integrating (4.5) from $t_i + T$ to $t^* + T$ yields

$$J_T(x_{t^*}, \widehat{u}) \leq J_T^*(x_{t_i}) - \int_{t_i}^{t^*} F(x^*(t'), u_T^*(t'))\, dt'. \tag{A.7}$$

Since $\widehat{u}$ is a feasible, but not necessarily optimal, solution

$$J_T^*(x_{t^*}) \leq J_T(x_{t^*}, \widehat{u}) \leq J_T^*(x_{t_i}) - \int_{t_i}^{t^*} F(x^*(t'), u_T^*(t'))\, dt'.$$

This completes the proof of the lemma. $\qquad\square$

Using these results, asymptotic stability as stated in Theorem 4.5 can now be proven similar to (Chen, 1997, Theorem 3.1).

Proof of Theorem 4.5. In the following, first stability of the closed-loop is proven. Given $\varepsilon > 0$, assume without loss of generality that (A.3) in Lemma A.3 holds for all states in the neighborhood of the origin defined by $\|x_t\|_\tau < \varepsilon$ and define $\beta = \underline{\alpha}_J(\varepsilon)$. Because of the continuity of $J_T^*(x_t)$ at $x_t = 0$, it is possible to find a $\delta_\varepsilon > 0$ such that $J_T^*(x_t) < \beta$

for all $\|x_t\|_\tau < \delta_\varepsilon$. Due to Lemma A.5, the optimal value functional $J_T^*(x_t)$ satisfies along trajectories of the closed-loop for all $t^* > t$

$$J_T^*(x_{t^*}) \leq J_T^*(x_t) - \int_t^{t^*} \underline{\alpha}_F(|x(t')|)\, dt'. \tag{A.8}$$

Hence, it is non-increasing along trajectories of the closed-loop and, therefore, for all $t^* > t$

$$\|x_t\|_\tau < \delta_\varepsilon \quad \Rightarrow \quad J_T^*(x_t) < \beta \quad \Rightarrow \quad J_T^*(x_{t^*}) < \beta \quad \Rightarrow \quad \|x_{t^*}\|_\tau < \varepsilon.$$

Thus, $x_t = 0$ is stable. In order to show asymptotic stability, use (A.8) iteratively to obtain

$$J_T^*(x_\infty) \leq J_T^*(x_t) - \int_t^\infty \underline{\alpha}_F(|x(t')|)\, dt'.$$

Since $J_T^*(x_t)$ is finite and $J_T^*(x_\infty) \geq 0$, the integral exists and is bounded, i.e.,

$$\int_t^\infty \underline{\alpha}_F(|x(t')|)\, dt' < \infty. \tag{A.9}$$

Moreover, $\|x_t\|_\tau$ is bounded for all time because the closed-loop is stable. With the input constraint set $\mathbb{U}$ compact and f continuous, it follows that $f(x(t), x(t-\tau), u(t))$ is bounded for all $t \in \mathbb{R}_{>0}$. Hence, $x(t)$ is uniformly continuous in t, which implies $|x(t)| \to 0$ for $t \to \infty$ according to Barbalat's Lemma (Barbalat, 1959; Khalil, 2002). $\qquad\square$

A.2 Proof of Theorem 4.10

The proof consists of two parts. In the first part, we show that the Lyapunov Inequality (4.5) in Assumption 4.6 holds. In the second part, we use this result in order to prove controlled positive invariance of Ω defined in (4.20).

Lyapunov Inequality (4.5)

Applying the Schur complement to the lower right block in (4.18), and pre- and post-multiplying by $\mathcal{P} = \begin{bmatrix} P & 0 \\ 0 & P \end{bmatrix}$, in which $P = \Lambda^{-1}$, one obtains

$$\begin{bmatrix} A_k^T P + P A_k + S + Q + K^T R K & P A_\tau \\ A_\tau^T P & -S \end{bmatrix} + \varepsilon \mathcal{P}^2 \prec 0. \tag{A.10}$$

The derivative of the cost functional E (4.19) along solutions of the closed-loop consisting of system (4.1) in combination with controller $u(t) = Kx(t)$ is

$$\begin{aligned} \dot{E}(x_t) = {} & x^T(t)\left[A_k^T P + P A_k\right]x(t) + 2x^T(t)P A_\tau x(t-\tau) \\ & + x^T(t)Sx(t) - x^T(t-\tau)Sx(t-\tau) + 2x^T(t)P\Phi(x_t, Kx(t)), \end{aligned} \tag{A.11}$$

in which $A_k = A + BK$. Comparing the results of (A.10) and (A.11), it is clear that $\dot{E}(x_t) \leq -F(x(t), Kx(t))$ holds if

$$2x^T(t)P\Phi(x_t, Kx(t)) \leq \varepsilon\, \lambda_{\min}^2(P)\, \left|(x^T(t), x^T(t-\tau))^T\right|^2 .$$

In order to show that this relation holds in the terminal region Ω defined by (4.20), note that property (4.8) is satisfied for all $\|x_t\|_\tau \leq \delta_\gamma$ with γ in (4.21). Therefore, the following holds

$$
\begin{aligned}
2x^T(t)P\Phi(x_t, Kx(t)) &\leq 2\lambda_{max}(P)\, |x(t)|\, |\Phi(x_t, Kx(t))| \\
&\overset{(4.8)}{\leq} 2\lambda_{max}(P)\, |x(t)|\, \gamma\, \left|(x^T(t), x^T(t-\tau))^T\right| \\
&\overset{(4.21)}{\leq} 2\lambda_{max}(P)\, |x(t)|\, \varepsilon \frac{\lambda_{\min}^2(P)}{2\,\lambda_{max}(P)}\, \left|(x^T(t), x^T(t-\tau))^T\right| \\
&\leq \varepsilon\, \lambda_{\min}^2(P)\, \left|(x^T(t), x^T(t-\tau))^T\right|^2 .
\end{aligned}
$$

Hence, $\dot{E}(x_t) \leq -F(x(t), Kx(t))$ for all x_t for which $\|x_t\|_\tau \leq \delta_\gamma$ and, consequently, for all x_t in the terminal region Ω.

Controlled positive invariance of Ω

In this part, the positive invariance of Ω is shown analogue to the proof of Theorem 4.7. Without loss of generality assume that $x_{t_0} \in \Omega$ for an arbitrary time instant t_0. For the sake of contradiction, assume that Ω is not positively invariant. Since $x(t)$ is a continuous function of time, there exists a $t_1 > t_0$ for which $x_{t_1} \notin \Omega$ and $\|x_t\|_\tau < \frac{3\delta_\gamma}{4}$ for all $t \leq t_1$. Note that $\dot{E}(x_t) \leq 0$ for all x_t with $\|x_t\|_\tau \leq \delta_\gamma$ as shown in the first part of this proof. Hence, $E(x_{t_1}) \leq E(x_{t_0})$ and $\|x_{t_1}\|_\tau > \delta_\gamma/2$ because we assume $x_{t_1} \notin \Omega$. It follows that there is a time t_2 with $t_0 < t_2 \leq t_1$ for which

$$|x(t_2)| > \frac{\delta_\gamma}{2} , \tag{A.12}$$

and due to $\dot{E} < 0$

$$E(x_{t_2}) \leq E(x_{t_0}) . \tag{A.13}$$

On the other hand, closer inspection of the definition of the terminal cost functional in (4.19) reveals $E(x_t) \geq \lambda_{\min}(P)\, |x(t)|^2$. Therefore,

$$E(x_{t_2}) \geq \lambda_{\min}(P)\, |x(t_2)|^2 \overset{(A.12)}{>} \lambda_{\min}(P)\, \frac{\delta_\gamma^2}{4} .$$

Using this result and (A.13), it directly follows that $E(x_{t_0}) > \lambda_{\min}(P)\frac{\delta_\gamma^2}{4}$, which contradicts the assumption that $x_{t_0} \in \Omega$. Hence, the terminal region Ω is positively invariant.

Furthermore, because of the invariance of Ω and $\dot{E}(x_t) \leq -F(x(t), Kx(t))$, it directly follows that the control law $u(t) = Kx(t)$ locally asymptotically stabilizes the nonlinear time-delay system (4.1).

A.3 Proof of Lemma 4.13

The proof uses ideas given in De Souza and Li (1995). Since

$$x(t - \tau) = x(t) - \int_{-\tau}^{0} \dot{x}(t + \theta) d\theta$$

$$= x(t) - \int_{-\tau}^{0} \tilde{f}(x(t + \theta), x(t - \tau + \theta), Kx(t + \theta)) + \Phi(x_{t+\theta}, Kx(t + \theta)) \, d\theta,$$

(A.14)

in which $\tilde{f}$ and Φ are defined in (4.7) and (4.6), any solution of system (4.1) is also a solution of the system

$$\dot{\xi} = (A_k + A_\tau)\xi(t) + \Phi(\xi_t, K\xi(t))$$

$$- A_\tau \int_{-\tau}^{0} [A_k\xi(t + \theta) + A_\tau\xi(t - \tau + \theta) + \Phi(\xi_{t+\theta}, K\xi(t + \theta))] \, d\theta \qquad \text{(A.15a)}$$

$$\xi(\theta) = \psi(\theta), \forall \theta \in [-2\tau, 0] \qquad \text{(A.15b)}$$

in which the short hand $A_k = A + BK$ is used. Hence, if Ω is positively invariant for the latter system (A.15), then it is also positively invariant for the original system (4.1).

Define a Razumikhin function candidate $V_1(\xi(t)) = \xi^T(t)P\xi(t)$ with the symmetric positive definite matrix $P = \Lambda^{-1} \succ 0$. The time derivative of V_1 along trajectories of (A.15) is

$$\dot{V}_1(\xi) = \xi^T(t) \left[(A_k + A_\tau)^T P + P(A_k + A_\tau) \right] \xi(t) + 2\xi^T(t)P\Phi(\xi_t, K\xi(t)) + \sum_{i=1}^{3} \eta_i(\xi, t)$$

(A.16)

in which

$$\eta_1(\xi, t) = -2 \int_{-\tau}^{0} \xi^T(t)PA_\tau A_k\xi(t + \theta) \, d\theta, \qquad \text{(A.17a)}$$

$$\eta_2(\xi, t) = -2 \int_{-\tau}^{0} \xi^T(t)PA_\tau^2\xi(t - \tau + \theta) \, d\theta, \qquad \text{(A.17b)}$$

$$\eta_3(\xi, t) = -2 \int_{-\tau}^{0} \xi^T(t)PA_\tau\Phi(\xi_{t+\theta}, K\xi(t + \theta)) \, d\theta. \qquad \text{(A.17c)}$$

For the symmetric matrices $P_i = \Lambda^{-1}\Lambda_i\Lambda^{-1} \succ 0$, $i \in \{1, 2, 3\}$, Inequality (4.26b) yields $P_i - P \le 0$, $i \in \{1, 2\}$. Furthermore, we know that for any $v, w \in \mathbb{R}^n$ and for any symmetric positive definite matrix $P_i \in \mathbb{R}^{n \times n}$

$$-2v^T w \le v^T P_i^{-1} v + w^T P_i w.$$

Motivated by Razumikhin-type arguments assume that

$$V_1(\xi(t + \theta)) \le V_1(\xi(t)), \forall \theta \in [-2\tau, 0]. \qquad \text{(A.18)}$$

Thus, it follows from using (A.17)–(A.18)

$$\eta_1(\xi, t) \leq \tau \xi^T(t) P A_\tau A_k P_1^{-1} A_k^T A_\tau^T P \xi(t) + \tau \xi^T(t) P \xi(t),$$
$$\eta_2(\xi, t) \leq \tau \xi^T(t) P A_\tau^2 P_2^{-1} (A_\tau^2)^T P \xi(t) + \tau \xi^T(t) P \xi(t),$$
$$\eta_3(\xi, t) \leq \tau \xi^T(t) P A_\tau P_3^{-1} A_\tau^T P \xi(t) + \int_{-\tau}^{0} \Phi(\xi_{t+\theta}, K\xi(t + \theta))^T P_3 \Phi(\xi_{t+\theta}, K\xi(t + \theta)) d\theta.$$

Substituting the result in (A.16) yields

$$\dot{V}_1(\xi) < \xi^T(t) \Theta \xi(t) + 2\xi^T(t) P \Phi(\xi_t, K\xi(t))$$
$$+ \int_{-\tau}^{0} \Phi(\xi_{t+\theta}, K\xi(t + \theta))^T P_3 \Phi(\xi_{t+\theta}, K\xi(t + \theta)) d\theta, \qquad (A.19)$$

in which

$$\Theta = \tau P A_\tau (A_k P_1^{-1} A_k^T + A_\tau P_2^{-1} A_\tau^T + P_3^{-1}) A_\tau^T P + 2\tau P + (A_k + A_\tau)^T P + P(A_k + A_\tau).$$

By using Equation (A.18), we know that $|\xi(t + \theta)| < \nu|\xi(t)|$ for all $\theta \in [-2\tau, 0]$ with $\nu^2 = \lambda_{\max}(P)/\lambda_{\min}(P)$. Using (4.8), we obtain

$$2\xi^T(t) P \Phi(\xi_t, K\xi(t)) \leq \underbrace{2\|P\| \gamma (1 + \nu)}_{\Sigma_1(\gamma)} |\xi(t)|^2 \qquad (A.20)$$

and

$$\int_{-\tau}^{0} \Phi(\xi_{t+\theta}, K\xi(t + \theta)^T) P_3 \Phi(\xi_{t+\theta}, K\xi(t + \theta)) d\theta \leq \underbrace{4 \tau \gamma^2 \nu^2 \|P_3\|}_{\Sigma_2(\gamma)} |\xi(t)|^2. \qquad (A.21)$$

Applying the Schur complement to (4.26a), using the substitutions $\Lambda = P^{-1}$, $\Lambda_i = P^{-1} P_i P^{-1}$ for $i \in \{1, 2, 3\}$, and $K = \Gamma \Lambda^{-1}$ and pre- and post-multiplying by P yields $\Theta = -W_1 \prec 0$. Now choose α in (4.23) small enough such that for all states $x_t \in \Omega$ the local control law satisfies the input constraints $u(t) = Kx(t) \in \mathbb{U}$ and Property (4.8) holds with γ small enough such that

$$\Sigma_1(\gamma) + \Sigma_2(\gamma) < \lambda_{\min}(W_1)/2. \qquad (A.22)$$

Using Equations (A.20)–(A.22) in combination with Equation (A.19), it can be ensured that $\dot{V}_1 < -\frac{\lambda_{\min}(W_1)}{2}|\xi(t)|^2$ whenever (A.18) holds. Note that (A.22) always holds for sufficiently small α because P is positive definite and (4.8). Thus, by Razumikhin-type arguments it follows that Ω is positively invariant, see Hale and Lunel (1993). This completes the proof.

A.4 Proof of Lemma 4.14

As shown in the first part of Appendix A.2, $\dot{E}(x_t) \leq -F(x(t), Kx(t))$ for all x_t with $\|x_t\|_\tau \leq \delta_\gamma$ if $\gamma \leq \varepsilon \frac{\lambda_{\min}^2(P)}{2 \lambda_{\max}(P)}$, compare (4.21) in Theorem 4.10. Hence, if α in the definition of Ω (4.27) is chosen small enough such that

$$\alpha < \lambda_{\min}(P) \delta_\gamma^2, \qquad (A.23)$$

then $\dot{E}(x_t) \leq -F(x(t), Kx(t))$ for all $x_t \in \Omega$.

A.5 Proof of Lemma 4.19

Pre- and post-multiplying the LMI (4.34) by $\mathcal{P} = \begin{bmatrix} P & 0 \\ 0 & P \end{bmatrix}$ yields

$$\begin{bmatrix} A_k^T P + P A_k + P + \tilde{\varepsilon} P^2 & P A_\tau \\ A_\tau^T P & -\frac{1}{\rho} P \end{bmatrix} \prec 0, \tag{A.24}$$

in which $A_k = A + BK$. Whenever (4.22) holds, i.e., $\forall \theta \in [-\tau, 0] : V(x(t+\theta)) \le \rho V(x(t))$, we obtain for the derivative of V along trajectories of the linear system (4.6)

$$\dot{V}(x(t)) = x(t)^T (PA_k + A_k P) x(t) + 2x(t)^T P A_\tau x(t - \tau)$$

$$\overset{(4.22)}{\le} x(t)^T (PA_k + A_k P) x(t) + 2x(t)^T P A_\tau x(t - \tau)$$

$$+ \underbrace{x(t)^T P x(t) - \frac{1}{\rho} x(t - \tau)^T P x(t - \tau)}_{\ge 0}$$

$$\overset{(A.24)}{\le} -\tilde{\varepsilon} x(t)^T P^2 x(t) \le -\tilde{\varepsilon} \lambda_{\min}(P^2) |x(t)|^2 .$$

This completes the proof.

A.6 Proof of Lemma 4.31

Due to fundamental properties of integrals and non-negativity of F, it follows that

$$\min_{t \in [t_1, t_2]} \left(F(t) + \int_{t-\tau}^{t} F(t') \, dt' \right) \le \frac{1}{t_2 - t_1} \int_{t_1}^{t_2} \left(F(t) + \int_{t-\tau}^{t} F(t') \, dt' \right) dt$$

$$\le \frac{1}{t_2 - t_1} \int_{t_1-\tau}^{t_2} F(t') \, dt' + \frac{1}{t_2 - t_1} \int_{t_1}^{t_2} \int_{t-\tau}^{t} F(t') \, dt' dt$$

$$\overset{(*)}{\le} \frac{1}{t_2 - t_1} \int_{t_1-\tau}^{t_2} F(t') dt' + \frac{1}{t_2 - t_1} \int_{t_1-\tau}^{t_2} \int_{t'}^{t'+\tau} F(t') dt dt'$$

$$= \frac{1}{t_2 - t_1} \int_{t_1-\tau}^{t_2} F(t') \, dt' + \frac{\tau}{t_2 - t_1} \int_{t_1-\tau}^{t_2} F(t') dt' .$$

For the interchange of the order of integration and the enlarged domain of integration in inequality $(*)$, see Figure A.1. The term $\int_{t_1}^{t_2} \int_{t-\tau}^{t} F(t') \, dt' dt$ results from integration over the domain within the black solid line, whereas $\int_{t_1-\tau}^{t_2} \int_{t'}^{t'+\tau} F(t') dt dt'$ corresponds to the larger domain additionally including the areas given by the gray dashed lines.

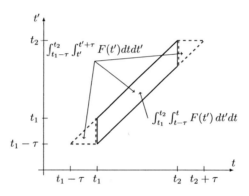

Figure A.1: Sketch for interchange of the order of integration in the proof of Lemma 4.31.

Bibliography

M. Alamir and G. Bornard. On the stability of receding horizon control of nonlinear discrete-time systems. *Systems & Control Letters*, 23(4):291–296, 1994.

M. Alamir and G. Bornard. Stability of a truncated infinite constrained receding horizon scheme: the general discrete nonlinear case. *Automatica*, 31(9):1353–1356, 1995.

N. Altmüller, L. Grüne, and K. Worthmann. Receding horizon optimal control for the wave equation. In *Proceedings of the 49th IEEE Conference on Decision and Control*, pages 3427–3432, Atlanta, GA, USA, 2010a.

N. Altmüller, L. Grüne, and K. Worthmann. Instantaneous control of the linear wave equation. In *Proceedings of the 19th International Symposium on Mathematical Theory of Networks and Systems*, pages 1895–1899, Budapest, Hungary, 2010b.

N. Altmüller, L. Grüne, and K. Worthmann. Improved stability estimates for MPC without terminal constraints applied to reaction diffusion PDEs. Preprint, 2012.

R. Amrit. *Optimizing Process Economics in Model Predictive Control*. PhD thesis, University of Wisconsin-Madison, 2011.

D. Angeli, R. Amrit, and J. B. Rawlings. On average performance and stability of economic model predictive control. *IEEE Transactions on Automatic Control*, 57(7):1615–1626, 2012.

C. Angrick. Nonlinear model predictive control of time-delay systems. Student thesis, University of Stuttgart, 2007.

M. Arcak. A global separation theorem for a new class of nonlinear observers. In *Proceedings of the 41st IEEE Conference on Decision and Control*, pages 676–681, Las Vegas, NV, USA, 2002.

A. N. Atassi and H. K. Khalil. A separation principle for the control of a class of nonlinear systems. *IEEE Transactions on Automatic Control*, 44(9):1672–1687, 1999.

A. N. Atassi and H. K. Khalil. Separation results for the stabilization of nonlinear systems using different high-gain observer designs. *Systems & Control Letters*, 39(3):183–191, 2000.

A. N. Atassi and H. K. Khalil. A separation principle for the control of a class of nonlinear systems. *IEEE Transactions on Automatic Control*, 46(5):742–746, 2001.

I. Barbalat. Systèmes d'équations différentielles d'oscillations nonlinéaires. *Rev. Roumaine Math. Pures Appl.*, 4:267–270, 1959.

R. E. Bellman. *Dynamic Programming*. Princeton University Press, New Jersey, NJ, USA, 1957.

A. Bemporad, F. Borrelli, and M. Morari. Model predictive control based on linear programming – the explicit solution. *IEEE Transactions on Automatic Control*, 47(12): 1974–1985, 2002a.

A. Bemporad, M. Morari, V. Dua, and E. N. Pistikopoulos. The explicit linear quadratic regulator for constrained systems. *Automatica*, 38(1):3–20, 2002b.

C. Böhm, T. Raff, M. Reble, and F. Allgöwer. LMI-based model predictive control for linear discrete-time periodic systems. In L. Magni, D. Raimondo, and F. Allgöwer, editors, *Nonlinear Model Predictive Control: Towards New Challenging Applications*, Lecture Notes in Control and Information Sciences, pages 99–108. Springer Verlag, 2009.

M. Boutayeb. Observers design for linear time-delay systems. *Systems & Control Letters*, 44(2):103–109, 2001.

R. W. Brockett. Asymptotic stability and feedback stabilization. In R. W. Brockett, R. S. Millman, and H. J. Sussmann, editors, *Differential Geometric Control Theory*, pages 181–191. Birkhäuser, Boston, 1983.

I. N. Bronstein, K. A. Semendjajew, G. Musiol, and H. Mühlig. *Taschenbuch der Mathematik*. Harri Deutsch, Frankfurt, Germany, 2000.

F. D. Brunner. Unconstrained model predictive control for nonlinear time-delay systems. Student thesis, University of Stuttgart, 2010.

E. F. Camacho and C. Bordons. *Model Predictive Control*. Springer, London, 2004.

C. C. Chen and L. Shaw. On receding horizon feedback control. *Automatica*, 18(3):349–352, 1982.

H. Chen. *Stability and Robustness Considerations in Nonlinear Model Predictive Control*. PhD thesis, University of Stuttgart, 1997.

H. Chen and F. Allgöwer. A quasi-infinite horizon nonlinear model predictive control scheme with guaranteed stability. *Automatica*, 34(10):1205–1218, 1998.

W.-H. Chen. Stability analysis of classic finite horizon model predictive control. *International Journal of Control, Automation, and Systems*, 8(2):187–197, 2010.

W.-H. Chen and Y. Cao. Stability analysis of constrained nonlinear model predictive control with terminal weighting. *Asian Journal of Control*, 14(6):1–8, 2012.

L. Chisci, A. Lombardi, and E. Mosca. Dual receding horizon control of constrained discrete-time systems. *European Journal of Control*, 2(4):278–285, 1996.

E. F. Costa and J. B. R. do Val. Stability of receding horizon control of nonlinear systems. In *Proceedings of the 42nd IEEE Conference on Decision and Control*, pages 2077–2081, Maui, HI, USA, 2003.

C. R. Cutler and B. L. Ramaker. Dynamic matrix control – a computer control algorithm. In *Proceedings of the Joint Automatic Control Conference*, pages 13–15, San Francisco, CA, USA, 1980.

M. L. Darby and M. Nikolaou. MPC: Current practice and challenges. *Control Engineering Practice*, 20(4):328–342, 2012.

G. De Nicolao, L. Magni, and R. Scattolini. Stabilizing receding-horizon control of nonlinear time-varying systems. *IEEE Transactions on Automatic Control*, 43(7):1030–1036, 1998.

C. E. De Souza and X. Li. Delay-dependent stability of linear time-delay systems: An LMI approach. In *Proceedings of the 3rd IEEE Mediterranean Symposium on Control and Automation*, pages 1–5, Limassol, Cyprus, 1995.

L. Del Re, F. Allgöwer, L. Glielmo, C. Guardiola, and I. Kolmanovsky, editors. *Automotive Model Predictive Control – Models, Methods and Applications*. Lecture Notes in Control and Information Sciences. Springer Verlag, Berlin, 2010.

M. Diehl, R. Findeisen, F. Allgöwer, H. G. Bock, and J. P. Schlöder. Nominal stability of real-time iteration scheme for nonlinear model predictive control. *IEE Proceedings Control Theory & Applications*, 152(3):296–308, 2004.

M. Diehl, R. Amrit, and J. B. Rawlings. A Lyapunov function for economic model predictive control. *IEEE Transactions on Automatic Control*, 56(3):703–707, 2011.

A. Domahidi, M. N. Zeilinger, M. Morari, and C. N. Jones. Learning a feasible and stabilizing explicit model predictive control law by robust optimization. In *Proceedings of the 50th IEEE Conference on Decision and Control and European Control Conference*, pages 513–519, Orlando, FL, USA, 2011.

S. Dubljevic, P. Mhaskar, N. H. El-Farra, and P. D. Christofides. Predictive control of transport-reaction processes. *Computers and Chemical Engineering*, 29(11–12):2335–2345, 2005.

S. Dubljevic, N. H. El-Farra, P. Mhaskar, and P. D. Christofides. Predictive control of parabolic PDEs with state and control constraints. *International Journal of Robust and Nonlinear Control*, 16(16):749–772, 2006a.

S. Dubljevic, P. Mhaskar, N. H. El-Farra, and P. D. Christofides. Predictive control of infinite dimensional systems. In *Proceedings of the 45th IEEE Conference on Decision and Control*, pages 93–100, San Diego, CA, USA, 2006b.

W. B. Dunbar and R. M. Murray. Distributed receding horizon control for multi-vehicle formation stabilization. *Automatica*, 42(4):549–558, 2006.

T. Faulwasser. Predictive path following without terminal constraints. In *Proceedings of the 20th International Symposium on Mathematical Theory of Networks and Systems*, Melbourne, VIC, Australia, 2012.

T. Faulwasser and R. Findeisen. Ein prädiktiver Ansatz zur Lösung nichtlinearer Pfadverfolgungsprobleme unter Beschränkungen. *at – Automatisierungstechnik*, 57(8):386–394, 2009a.

T. Faulwasser and R. Findeisen. Nonlinear model predictive path-following control. In L. Magni, D. Raimondo, and F. Allgöwer, editors, *Nonlinear Model Predictive Control: Towards New Challenging Applications*, Lecture Notes in Control and Information Sciences, pages 335–343. Springer Verlag, 2009b.

A. Ferramosca, D. Limon, I. Alvarado, T. Alamo, and E. F. Camacho. MPC for tracking with optimal closed-loop performance. *Automatica*, 45(8):1975–1978, 2009.

G. Ferrari-Trecate, L. Galbusera, M. P. E. Marciandi, and R. Scattolini. Model predictive control schemes for consensus in multi-agent systems with single- and double-integrator dynamics. *IEEE Transactions on Automatic Control*, 54(11):2560–2572, 2009.

R. Findeisen. *Nonlinear Model Predictive Control: A Sampled-Data Feedback Perspective*. PhD thesis, University of Stuttgart, 2004.

R. Findeisen and F. Allgöwer. A nonlinear model predictive control scheme for the stabilization of setpoint families. *Journal A, Benelux Quarterly Journal on Automatic Control*, 41(1):37–45, 2000.

R. Findeisen and P. Varutti. Stabilizing nonlinear predictive control over nondeterministic communication networks. In L. Magni, D. Raimondo, and F. Allgöwer, editors, *Nonlinear Model Predictive Control: Towards New Challenging Applications*, Lecture Notes in Control and Information Sciences, pages 167–179. Springer Verlag, 2009.

R. Findeisen, L. Imsland, F. Allgöwer, and B. A. Foss. State and output feedback nonlinear model predictive control: An overview. *European Journal of Control*, 9(2–3):179–195, 2003.

F. A. C. C. Fontes. A general framework to design stabilizing nonlinear model predictive controllers. *Systems & Control Letters*, 42(2):127–143, 2001.

F. A. C. C. Fontes. Discontinuous feedbacks, discontinuous optimal controls, and continuous-time model predictive control. *International Journal of Robust and Nonlinear Control*, 13(3–4):191–209, 2003.

A. Freuer, M. Reble, C. Böhm, and F. Allgöwer. Efficient model predictive control for linear periodic systems. In *Proceedings of the 19th International Symposium on Mathematical Theory of Networks and Systems*, pages 1403–1409, Budapest, Hungary, 2010.

D. Georges. Infinite-dimensional nonlinear predictive control design for open-channel hydraulic systems. *Network and Heterogeneous Media*, 2(4):1–18, 2009.

A. Germani, C. Manes, and P. Pepe. A state observer for nonlinear delay systems. In *Proceedings of the 37th IEEE Conference on Decision and Control*, pages 355–360, Tampa, CA, USA, 1998.

A. Germani, C. Manes, and P. Pepe. An observer for M.I.M.O. nonlinear delay systems. In *Proceedings of the 14th IFAC World Congress*, pages 243–248, Beijing, China, 1999.

A. Germani, C. Manes, and P. Pepe. Local asymptotic stability for nonlinear state feedback delay systems. *Kybernetika*, 36(1):31–42, 2000.

A. Germani, C. Manes, and P. Pepe. An asymptotic state observer for a class of nonlinear delay systems. *Kybernetika*, 37(4):459–478, 2001.

A. Germani, C. Manes, and P. Pepe. Separation theorems for a class of retarded nonlinear systems. In *Proceedings of the IFAC Workshop on Time-Delay Systems*, Prague, Czech Republic, 2010.

P. Giselsson. Adaptive nonlinear model predictive control with suboptimality and stability guarantees. In *Proceedings of the 49th IEEE Conference on Decision and Control*, pages 3644–3649, Atlanta, GA, USA, 2010.

R. Gondhalekar and C. N. Jones. MPC of constrained discrete-time linear periodic systems – a framework for asynchronous control: Strong feasibility, stability and optimality via periodic invariance. *Automatica*, 47(2):326–333, 2011.

G. C. Goodwin, M. M. Seron, and J. A. De Doná. *Constrained Control and Estimation: An Optimisation Approach*. Springer, London, 2005.

K. Graichen. A fixed-point iteration scheme for real-time model predictive control. *Automatica*, 48(7):1300–1305, 2012.

K. Graichen and A. Kugi. Stability and incremental improvement of suboptimal MPC without terminal constraints. *IEEE Transactions on Automatic Control*, 55(11):2576–2580, 2010.

K. Graichen, M. Egretzberger, and A. Kugi. Ein suboptimaler Ansatz zur schnellen modellprädiktiven Regelung nichtlinearer Systeme (A suboptimal approach to real-time model predictive control of nonlinear systems). *at – Automatisierungstechnik*, 58(8): 447–456, 2010.

G. Grimm, M. J. Messina, S. E. Tuna, and A. R. Teel. Model predictive control: for want of a local control Lyapunov function, all is not lost. *IEEE Transactions on Automatic Control*, 50(5):546–558, 2005.

L. Grüne. Computing stability and performance bounds for unconstrained NMPC schemes. In *Proceedings of the 46th IEEE Conference on Decision and Control*, pages 1263–1268, New Orleans, LA, USA, 2007.

L. Grüne. Analysis and design of unconstrained nonlinear MPC schemes for finite and infinite dimensional systems. *SIAM Journal on Control and Optimization*, 48(2):1206–1228, 2009.

L. Grüne. Optimal invariance via receding horizon control. In *Proceedings of the 50th IEEE Conference on Decision and Control and European Control Conference*, pages 2668–2673, Orlando, FL, USA, 2011.

L. Grüne. Economic MPC without terminal constraints. In *Proceedings of the 20th International Symposium on Mathematical Theory of Networks and Systems*, Melbourne, VIC, Australia, 2012.

L. Grüne. Economic receding horizon control without terminal constraints. *Automatica*, 49 (3):725–734, 2013.

L. Grüne and J. Pannek. Trajectory based suboptimality estimates for receding horizon controllers. In *Proceedings of the 18th International Symposium on Mathematical Theory of Networks and Systems*, Blacksburg, VA, USA, 2008.

L. Grüne and J. Pannek. Practical NMPC suboptimality estimates along trajectories. *Systems & Control Letters*, 58(3):161–168, 2009.

L. Grüne and J. Pannek. Analysis of unconstrained NMPC schemes with incomplete optimization. In *Proceedings of the 8th IFAC Symposium on Nonlinear Control Systems*, pages 238–243, Bologna, Italy, 2010.

L. Grüne and J. Pannek. *Nonlinear Model Predictive Control: Theory and Algorithms*. Springer, London, 2011.

L. Grüne and A. Rantzer. Suboptimality estimates for receding horizon controllers. In *Proceedings of the 17th International Symposium on Mathematical Theory of Networks and Systems*, pages 120–127, Kyoto, Japan, 2006.

L. Grüne and A. Rantzer. On the infinite horizon performance of receding horizon controllers. *IEEE Transactions on Automatic Control*, 53(9):2100–2111, 2008.

L. Grüne and K. Worthmann. A distributed NMPC scheme without stabilizing terminal constraints. In R. Johansson and A. Rantzer, editors, *Distributed Decision Making and Control*, Lecture Notes in Control and Information Sciences, pages 259–285. Springer Verlag, 2012.

L. Grüne, J. Pannek, and K. Worthmann. A networked unconstrained nonlinear MPC scheme. In *Proceedings of the European Control Conference*, pages 371–376, Budapest, Hungary, 2009a.

L. Grüne, J. Pannek, and K. Worthmann. A prediction based control scheme for networked systems with delays and packet dropouts. In *Proceedings of the 48th IEEE Conference on Decision and Control and 28th Chinese Control Conference*, pages 537–542, Shanghai, China, 2009b.

L. Grüne, J. Pannek, M. Seehafer, and K. Worthmann. Analysis of unconstrained nonlinear MPC schemes with time-varying control horizon. *SIAM Journal on Control and Optimization*, 48(8):4938–4962, 2010a.

L. Grüne, M. von Lossow, J. Pannek, and K. Worthmann. MPC: implications of a growth condition on exponentially controllable systems. In *Proceedings of the 8th IFAC Symposium on Nonlinear Control Systems*, pages 385–390, Bologna, Italy, 2010b.

L. Grüne, M. von Lossow, and K. Worthmann. NMPC suboptimality estimates for sampled-data continuous systems. In M. Diehl, F. Glineur, E. Jarlebring, and W. Michiels, editors, *Recent Trends in Optimization and its Applications in Engineering*, pages 329–338. Springer Verlag, 2010c.

L. Grüne, J. Pannek, and K. Worthmann. Ensuring stability in networked systems with nonlinear MPC for continuous time systems. In *Proceedings of the 51st IEEE Conference on Decision and Control*, pages 14–19, Maui, HI, USA, 2012.

K. Gu, V. L. Kharitonov, and J. Chen. *Stability of Time-Delay Systems*. Birkhäuser, Boston, MA, USA, 2003.

J. K. Hale. *Theory of Functional Differential Equations*. Springer Verlag, New York, NY, USA, 1977.

J. K. Hale and S. M. V. Lunel. *Introduction to Functional Differential Equations*. Springer Verlag, New York, NY, USA, 1993.

W. Halter. Application aspects of different NMPC stability methods for fixed-wing UAVs. Student thesis, University of Stuttgart, 2012.

C. Han, X. Liu, and H. Zhang. Robust model predictive control for continuous uncertain systems with state delay. *Journal of Control Theory and Applications*, 6:189–194, 2008.

J. Hasenauer. Personal communication, 2012.

D.-F. He, L. Yu, and X.-L. Song. Optimization-based stabilization of constrained nonlinear systems: A receding horizon approach. In *Proceedings of the 18th IFAC World Congress*, pages 4904–4908, Milan, Italy, 2011.

Y. Q. He and J. D. Han. Nonlinear model predictive control with regulable computational cost. *Asian Journal of Control*, 14(3):1–8, 2010.

B. Hu and A. Linnemann. Toward infinite-horizon optimality in nonlinear model predictive control. *IEEE Transactions on Automatic Control*, 47(4):679–682, 2002.

X.-B. Hu and W.-H. Chen. Model predictive control for constrained systems with uncertain state-delays. *International Journal of Robust and Nonlinear Control*, 14(17):1421–1432, 2004.

C.-C. Hua, Q.-G. Wang, and X.-P. Guan. Memoryless state feedback controller design for time delay systems with matched uncertain nonlinearities. *IEEE Transactions on Automatic Control*, 53(3):801–807, 2008.

R. Huang, E. Harinath, and L. T. Biegler. Lyapunov stability of economically-oriented NMPC for cyclic processes. *Journal of Process Control*, 21(4):501–509, 2011.

J. M. Igreja, J. M. Lemos, and S. J. Costa. Robust pointwise min-norm control of distributed systems with fluid flow. In *Proceedings of the 50th IEEE Conference on Decision and Control and European Control Conference*, pages 266–2667, Orlando, FL, USA, 2011.

K. Ito and K. Kunisch. Receding horizon optimal control for infinite dimensional systems. *ESAIM: Control, Optimisation and Calculus of Variations*, 8:741–760, 2002.

A. Jadbabaie. *Receding Horizon Control of Nonlinear Systems: A Control Lyapunov Function Approach*. PhD thesis, California Institute of Technology, 2000.

A. Jadbabaie and J. Hauser. On the stability of receding horizon control with a general terminal cost. *IEEE Transactions on Automatic Control*, 50(5):674–678, 2005.

A. Jadbabaie, J. Yu, and J. Hauser. Stabilizing receding horizon control of nonlinear systems: A control Lyapunov function approach. In *Proceedings of the American Control Conference*, pages 1535–1539, San Diego, CA, USA, 1999.

A. Jadbabaie, J. Primbs, and J. Hauser. Unconstrained receding horizon control with no terminal cost. In *Proceedings of the American Control Conference*, pages 3055–3060, Arlington, VA, USA, 2001a.

A. Jadbabaie, J. Yu, and J. Hauser. Unconstrained receding-horizon control of nonlinear systems. *IEEE Transactions on Automatic Control*, 46(5):776–783, 2001b.

M. Jankovic. Control Lyapunov-Razumikhin functions and robust stabilization of time delay systems. *IEEE Transactions on Automatic Control*, 46(7):1048–1060, 2001.

M. Jankovic. Control of nonlinear systems with time-delay. In *Proceedings of the 42nd IEEE Conference on Decision and Control*, pages 4545–4550, Maui, HI, USA, 2003.

M. Jankovic. Stabilization of nonlinear time delay systems with delay-independent feedback. In *Proceedings of the American Control Conference*, pages 4253–4258, Portland, OR, USA, 2005.

S. C. Jeong and P. Park. Constrained MPC algorithm for uncertain time-varying systems with state-delay. *IEEE Transactions on Automatic Control*, 50(2):257–263, 2005.

T. A. Johansen. Approximate explicit receding horizon control of constrained nonlinear systems. *Automatica*, 40(4):293–300, 2004.

C. N. Jones and M. Morari. Polytopic approximation of explicit model predictive controllers. *IEEE Transactions on Automatic Control*, 55(11):2542–2553, 2010.

C. N. Jones, M. Baric, and M. Morari. Multiparametric linear programming with applications to control. *European Journal of Control*, 13(2-3):152–170, 2007.

M. Kano and M. Ogawa. The state of the art in chemical process control in Japan: Good practice and questionnaire survey. *Journal of Process Control*, 20(9):969–982, 2010.

S. S. Keerthi and E. G. Gilbert. Optimal infinite-horizon feedback laws for a general class of constrained discrete-time systems: Stability and moving-horizon approximations. *Journal of Optimization Theory and Applications*, 57(2):265–293, 1988.

T. Keviczky, F. Borrelli, and G. J. Balas. Decentralized receding horizon control for large scale dynamically decoupled systems. *Automatica*, 42(12):2105–2115, 2006.

H. K. Khalil. *Nonlinear Systems*. Prentice-Hall, Upper Saddle River, NJ, USA, 3rd edition, 2002.

V. L. Kharitonov and A. P. Zhabko. Lyapunov-Krasovskii approach to the robust stability analysis of time-delay systems. *Automatica*, 39(1):15–20, 2003.

V. Kolmanovskii and A. Myshkis. *Introduction to the Theory and Applications of Functional Differential Equations.* Kluwer Academic Publishers, Dordrecht, The Netherlands, 1999.

A. J. Koshkouei and K. J. Burnham. Discontinuous observers for non-linear time-delay systems. *International Journal of Systems Science*, 40(4):383–392, 2009.

S. L. O. Kothare and M. Morari. Contractive model predictive control for constrained nonlinear systems. *IEEE Transactions on Automatic Control*, 45(6):1053–1071, 2000.

S. Küchler and O. Sawodny. Nonlinear control of an active heave compensation system with time-delay. In *Proceedings of the IEEE Conference Control Applications*, pages 1313–1318, 2010.

M. Kvasnica, J. Löfberg, and M. Fikar. Stabilizing polynomial approximation of explicit MPC. *Automatica*, 47(10):2292–2297, 2011.

W. H. Kwon, Y. S. Lee, and S. H. Han. Receding horizon predictive control for non-linear time-delay systems. In *International Conference on Control, Automation and Systems*, pages 107–111, Cheju National Univ. Jeju, Korea, 2001a.

W. H. Kwon, Y. S. Lee, and S. H. Han. Receding horizon predictive control for nonlinear time-delay systems with and without input constraints. In *Proceedings of the 6th IFAC Symposium on Dynamics and Control of Process Systems*, pages 277–282, Jejudo Island, Korea, 2001b.

W. H. Kwon, Y. S. Lee, and S. H. Han. Receding horizon predictive control for linear time-delay systems. *SICE Annual Conference*, 2:1377–1382, 2003.

W. H. Kwon, Y. S. Lee, and S. H. Han. General receding horizon control for linear time-delay systems. *Automatica*, 40(9):1603–1611, 2004.

W. Langson, I. Chryssochoos, S. V. Raković, and D. Q. Mayne. Robust model predictive control using tubes. *Automatica*, 40(1):125–133, 2004.

M. Lazar. *Model Predictive Control of Hybrid Systems: Stability and Robustness.* PhD thesis, Technische Universitiet Eindhoven, 2006.

E. B. Lee and L. Markus. *Foundations of Optimal Control Theory.* John Wiley & Sons, New York, NY, USA, 1967.

S. M. Lee, S. C. Jeong, and S. C. Won. Robust model predictive control for LPV systems with delayed state using relaxation matrices. In *Proceedings of the American Control Conference*, pages 716–721, San Francisco, CA, USA, 2011.

D. Li and Y. Xi. Constrained feedback robust model predictive control for polytopic uncertain systems with time delays. *International Journal of Systems Science*, 42(10): 1651–1660, 2011.

D. Limon, T. Alamo, and E. F. Camacho. Stable constrained MPC without terminal constraint. In *Proceedings of the American Control Conference*, pages 4893–4898, Denver, CO, USA, 2003.

D. Limon, T. Alamo, F. Salas, and E. F. Camacho. On the stability of constrained MPC without terminal constraint. *IEEE Transactions on Automatic Control*, 51(5):832–836, 2006.

D. Limon, I. Alvarado, T. Alamo, and E. F. Camacho. MPC for tracking piecewise constant references for constrained linear systems. *Automatica*, 44(9):2382–2387, 2008.

B. Lincoln and A. Rantzer. Relaxing dynamic programming. *IEEE Transactions on Automatic Control*, 51(8):1249–1260, 2006.

J. L. Löfberg. YALMIP : A toolbox for modeling and optimization in MATLAB. In *Proceedings of the CACSD Conference*, Taipei, Taiwan, 2004.

M.-C. Lu. Adaptive receding horizon control for a class of nonlinear differential difference systems. In *Proceedings of the 50th IEEE Conference on Decision and Control and European Control Conference*, pages 1116–1121, Orlando, FL, USA, 2011.

J. M. Maciejowski. *Predictive Control with Constraints*. Prentice-Hall, Englewood Cliffs, NJ, USA, 2002.

L. Magni and R. Scattolini. Stabilizing model predictive control of nonlinear continuous time systems. *Annual Reviews in Control*, 28(1):1–11, 2004.

L. Magni, G. De Nicolao, L. Magnani, and R. Scattolini. A stabilizing model-based predictive control algorithm for nonlinear systems. *Automatica*, 37(9):1351–1362, 2001.

R. Mahboobi Esfanjani and S. K. Y. Nikravesh. Stabilising predictive control of non-linear time-delay systems using control Lyapunov-Krasovskii functionals. *IET Control Theory & Applications*, 3(10):1395–1400, 2009a.

R. Mahboobi Esfanjani and S. K. Y. Nikravesh. Robust model predictive control for constrained distributed delay systems. In *XXII International Symposium on Information, Communication and Automation Technologies*, Sarajevo, Bosnia Herzegovina, 2009b.

R. Mahboobi Esfanjani and S. K. Y. Nikravesh. Predictive control for a class of distributed delay systems using Chebyshev polynomials. *International Journal of Computer Mathematics*, 87(7):1591–1601, 2010.

R. Mahboobi Esfanjani and S. K. Y. Nikravesh. Stabilizing model predictive control for constrained nonlinear distributed delay systems. *ISA Transactions*, 50(2):201–206, 2011.

R. Mahboobi Esfanjani, M. Reble, U. Münz, S. K. Y. Nikravesh, and F. Allgöwer. Model predictive control of constrained nonlinear time-delay systems. In *Proceedings of the 48th IEEE Conference on Decision and Control and 28th Chinese Control Conference*, pages 1324–1329, Shanghai, China, 2009.

L. A. Márquez-Martínez and C. H. Moog. Input-output feedback linearization of time-delay systems. *IEEE Transactions on Automatic Control*, 49(5):781–785, 2004.

D. Q. Mayne and H. Michalska. Receding horizon control of nonlinear systems. In *Proceedings of the 28th IEEE Conference on Decision and Control*, pages 107–108, Tampa, FL, USA, 1989.

D. Q. Mayne and H. Michalska. Receding horizon control of nonlinear systems. *IEEE Transactions on Automatic Control*, 35(7):814–824, 1990.

D. Q. Mayne, J. B. Rawlings, C. V. Rao, and P. O. M. Scokaert. Constrained model predictive control: stability and optimality. *Automatica*, 26(6):789–814, 2000.

D. Q. Mayne, M. M. Seron, and S. V. Raković. Robust model predictive control of constrained linear systems with bounded disturbances. *Automatica*, 41(2):219–224, 2005.

F. Mazenc and P.-A. Bliman. Backstepping design for time-delay nonlinear systems. *IEEE Transactions on Automatic Control*, 51(1):149–154, 2006.

J. S. Mejía and D. M. Stipanović. A modified contractive model predictive control approach. In *Proceedings of the 48th IEEE Conference on Decision and Control and 28th Chinese Control Conference*, pages 1968–1973, Shanghai, China, 2009.

D. Melchor-Aguilar and S.-I. Niculescu. Estimates of the attraction region for a class of nonlinear time-delay systems. *IMA Journal of Mathematical Control and Information*, 24:523–550, 2007.

M. J. Messina. *Model predictive control of constrained discrete-time nonlinear systems: stability and robustness*. PhD thesis, University of California, Santa Barbara, 2006.

P. Mhaskar, N. H. El-Farra, and P. D. Christofides. Stabilization of nonlinear systems with state and control constraints using Lyapunov-based predictive control. *Systems & Control Letters*, 55(8):650–659, 2006.

H. Michalska. Receding horizon stabilizing control without terminal constraint on the state. In *Proceedings of the 35th IEEE Conference on Decision and Control*, pages 2608–2613, Kobe, Japan, 1996.

H. Michalska and D. Q. Mayne. Receding horizon control of nonlinear systems without differentiability of the optimal value function. *Systems & Control Letters*, 16(2):123–130, 1991.

H. Michalska and D. Q. Mayne. Robust receding horizon control of constrained nonlinear systems. *IEEE Transactions on Automatic Control*, 38(11):1623–1633, 1993.

L. Mohammadi, S. Dubljevic, and J. F. Forbes. Robust characteristic-based MPC of a fixed-bed reactor. In *Proceedings of the American Control Conference*, pages 4421–4426, Baltimore, MD, USA, 2010.

D. Muñoz de la Peña and P. D. Christofides. Lyapunov-based model predictive control of nonlinear systems subject to data losses. *IEEE Transactions on Automatic Control*, 53 (9):2076–2089, 2008.

M. A. Müller, M. Reble, and F. Allgöwer. A general distributed MPC framework for cooperative control. In *Proceedings of the 18th IFAC World Congress*, pages 7987–7992, Milan, Italy, 2011.

M. A. Müller, M. Reble, and F. Allgöwer. Cooperative control of dynamically decoupled systems via distributed model predictive control. *International Journal of Robust and Nonlinear Control*, 22(12):1376–1397, 2012.

V. Nevistić and J. A. Primbs. Receding horizon quadratic optimal control: Performance bounds for a finite horizon strategy. In *Proceedings of the European Control Conference*, Brussels, Belgium, 1997.

H. A. Nour Eldin. *Optimierung linearer Regelsysteme mit quadratischer Zielfunktion.* Lecture Notes in Operations Research and Mathematical Systems. Springer Verlag, Berlin, 1971.

T. Oguchi, A. Watanabe, and N. T. Input-output linearization of retarded non-linear systems by using an extension of Lie derivative. *International Journal of Control*, 75(8): 582–590, 2002.

K. Ohsumi and T. Ohtsuka. Nonlinear receding horizon control of probability density functions. In *Proceedings of the 8th IFAC Symposium on Nonlinear Control Systems*, pages 735–740, Bologna, Italy, 2010.

K. Ohsumi and T. Ohtsuka. Particle model predictive control for probability density functions. In *Proceedings of the 18th IFAC World Congress*, pages 7993–7998, Milan, Italy, 2011.

T. Ohtsuka. A continuation/GMRES method for fast computation of nonlinear receding horizon control. *Automatica*, 40(4):563–574, 2004.

Y. Ou and E. Schuster. On the stability of receding horizon control of bilinear parabolic PDE systems. In *Proceedings of the 49th IEEE Conference on Decision and Control*, pages 851–857, Atlanta, GA, USA, 2010.

J. Pannek. *Receding Horizon Control: A Suboptimality–based Approach.* PhD thesis, University of Bayreuth, 2009.

G. Pannocchia, J. B. Rawlings, and S. J. Wright. Conditions under which suboptimal nonlinear MPC is inherently robust. *Systems & Control Letters*, 60(9):747–755, 2011.

A. Papachristodoulou. Analysis of nonlinear time-delay systems using the sum of squares decomposition. In *Proceedings of the American Control Conference*, pages 4153–4158, Boston, MA, USA, 2004.

A. Papachristodoulou. Robust stabilization of nonlinear time delay systems using convex optimization. In *Proceedings of the 44th IEEE Conference on Decision and Control and European Control Conference*, pages 5788–5793, Seville, Spain, 2005.

A. Papachristodoulou, M. Peet, and S. Lall. Constructing Lyapunov-Krasovskii functionals for linear time delay systems. In *Proceedings of the American Control Conference*, pages 2845–2850, Portland, OR, USA, 2005.

T. Parisini and R. Zoppoli. A receding-horizon regulator for nonlinear systems and a neural approximation. *Automatica*, 31(10):1443–1451, 1995.

T. V. Pham, D. Georges, and G. Besançon. Infinite-dimensional receding horizon optimal control for an open-channel system. In *Proceedings of the 8th IFAC Symposium on Nonlinear Control Systems*, pages 391–396, Bologna, Italy, 2010a.

T. V. Pham, D. Georges, and G. Besançon. On the use of a global control Lyapunov functional in infinite-dimensional predictive control. In *Proceedings of the 4th IFAC Symposium on System, Structure and Control*, pages 363–369, Ancona, Italy, 2010b.

T. V. Pham, D. Georges, and G. Besançon. Predictive control with guaranteed stability for hyperbolic systems of conservation laws. In *Proceedings of the 49th IEEE Conference on Decision and Control*, pages 6932–6937, Atlanta, GA, USA, 2010c.

T. V. Pham, D. Georges, and G. Besançon. Receding optimal boundary control of non-linear hyperbolic systems of conservation laws. In *Proceedings of the 18th IFAC World Congress*, pages 8601–8606, Milan, Italy, 2011.

T. V. Pham, D. Georges, and G. Besançon. Predictive control with terminal constraint for 2x2 hyperbolic systems of conservation laws. In *Proceedings of the 51st IEEE Conference on Decision and Control*, pages 6412–6417, Maui, HI, USA, 2012.

G. Pin and T. Parisini. Networked predictive control of uncertain constrained nonlinear systems: Recursive feasibility and input-to-state stability analysis. *IEEE Transactions on Automatic Control*, 56(1):72–87, 2011.

E. Polak and T. H. Yang. Moving horizon control of linear systems with input saturation, disturbances and plant uncertainty – Part 1. robustness. *International Journal of Control*, 58(3):613–638, 1993a.

E. Polak and T. H. Yang. Moving horizon control of linear systems with input saturation, disturbances and plant uncertainty – Part 2. disturbance rejection and tracking. *International Journal of Control*, 58(3):639–663, 1993b.

A. I. Propoi. Use of linear programming methods for synthesizing sampled data automatic systems. *Automation and Remote Control*, 24(7):837–844, 1963.

S. J. Qin and T. A. Badgwell. An overview of nonlinear model predictive control applications. In A. Zheng and F. Allgöwer, editors, *Nonlinear Predictive Control*, pages 369–392. Birkhäuser, 2000.

S. J. Qin and T. A. Badgwell. A survey of industrial model predictive control technology. *Control Engineering Practice*, 11(7):733–764, 2003.

D. E. Quevedo, J. Østergaard, and D. Nešić. Packetized predictive control of stochastic systems over bit-rate limited channels with packet loss. *IEEE Transactions on Automatic Control*, 56(12):2854–2868, 2011.

M. D. Rafal and W. F. Stevens. Discrete dynamic optimization applied to on-line optimal control. *AIChE Journal*, 14(1):85–91, 1968.

T. Raff and F. Allgöwer. An EKF-based observer for nonlinear time-delay systems. In *Proceedings of the American Control Conference*, pages 3130–3133, Minneapolis, MN, USA, 2006.

T. Raff, S. Huber, Z. K. Nagy, and F. Allgöwer. Nonlinear model predictive control of a four tank system: An experimental stability study. In *Proceedings of the IEEE Conference Control Applications*, pages 237–242, Munich, Germany, 2006.

T. Raff, C. Angrick, R. Findeisen, J. S. Kim, and F. Allgöwer. Model predictive control for nonlinear time-delay systems. In *Proceedings of the 7th IFAC Symposium on Nonlinear Systems*, Pretoria, South Africa, 2007.

D. M. Raimondo, D. Limon, M. Lazar, L. Magni, and E. Camacho. Min-max model predictive control of nonlinear systems: a unifying overview on stability. *European Journal of Control*, 15(1):5–21, 2009.

D. M. Raimondo, S. Riverso, C. N. Jones, and M. Morari. A robust explicit nonlinear MPC controller with input-to-state stability guarantees. In *Proceedings of the 18th IFAC World Congress*, pages 9284–9289, Milan, Italy, 2011.

J. B. Rawlings and D. Q. Mayne. *Model Predictive Control: Theory and Design*. Nob Hill Publishing, Madison, WI, USA, 2009.

M. Reble and F. Allgöwer. Stabilizing design parameters for model predictive control of constrained nonlinear time-delay systems. In *Proceedings of the IFAC Workshop on Time-Delay Systems*, Prague, Czech Republic, 2010a.

M. Reble and F. Allgöwer. General design parameters of model predictive control for nonlinear time-delay systems. In *Proceedings of the 49th IEEE Conference on Decision and Control*, pages 176–181, Atlanta, GA, USA, 2010b.

M. Reble and F. Allgöwer. Unconstrained nonlinear model predictive control and suboptimality estimates for continuous-time systems. In *Proceedings of the 18th IFAC World Congress*, pages 6733–6738, Milan, Italy, 2011.

M. Reble and F. Allgöwer. Design of terminal cost functionals and terminal regions for model predictive control of nonlinear time-delay systems. In R. Sipahi, T. Vyhlidal, P. Pepe, and S.-I. Niculescu, editors, *Time Delay Systems: Methods, Applications and New Trends*, Lecture Notes in Control and Information Sciences, pages 355–366. Springer Verlag, 2012a.

M. Reble and F. Allgöwer. Unconstrained model predictive control and suboptimality estimates for nonlinear continuous-time systems. *Automatica*, 48(8):1812–1817, 2012b.

M. Reble, C. Böhm, and F. Allgöwer. Nonlinear model predictive control for periodic systems using LMIs. In *Proceedings of the European Control Conference*, pages 3365–3370, Budapest, Hungary, 2009.

M. Reble, F. D. Brunner, and F. Allgöwer. Model predictive control for nonlinear time-delay systems without terminal constraint. In *Proceedings of the 18th IFAC World Congress*, pages 9254–9259, Milan, Italy, 2011a.

M. Reble, R. Mahboobi Esfanjani, S. K. Y. Nikravesh, and F. Allgöwer. Model predictive control of constrained nonlinear time-delay systems. *IMA Journal of Mathematical Control and Information*, 28(2):183–201, 2011b.

M. Reble, M. A. Müller, and F. Allgöwer. Unconstrained model predictive control and suboptimality estimates for nonlinear time-delay systems. In *Proceedings of the 50th IEEE Conference on Decision and Control and European Control Conference*, pages 7599–7604, Orlando, FL, USA, 2011c.

M. Reble, D. E. Quevedo, and F. Allgöwer. Stochastic stability and performance estimates of packetized unconstrained model predictive control for networked control systems. In *Proceedings of the 9th IEEE International Conference on Control & Automation*, pages 171–176, Santiago, Chile, 2011d.

M. Reble, D. E. Quevedo, and F. Allgöwer. A unifying framework for stability in MPC using a generalized integral terminal cost. In *Proceedings of the American Control Conference*, pages 1211–1216, Montréal, Canada, 2012a.

M. Reble, D. E. Quevedo, and F. Allgöwer. Improved stability conditions for unconstrained nonlinear model predictive control by using additional weighting terms. In *Proceedings of the 51st IEEE Conference on Decision and Control*, pages 2625–2630, Maui, HI, USA, 2012b.

M. Reble, D. E. Quevedo, and F. Allgöwer. Control over erasure channels: Stochastic stability and performance of packetized unconstrained model predictive control. *International Journal of Robust and Nonlinear Control*, 2012c. available online, DOI: 10.1002/rnc.2853.

J. Richalet, A. Rault, J. L. Testud, and J. Papon. Model predictive heuristic control: Applications to industrial processes. *Automatica*, 14(5):413–428, 1978.

J.-P. Richard. Time-delay systems: an overview of some recent advances and open problems. *Automatica*, 39(10):1667–1694, 2003.

B. J. P. Roset, M. Lazar, W. P. M. H. Heemels, and H. Nijmeijer. A stabilizing output based nonlinear model predictive control scheme. In *Proceedings of the 45th IEEE Conference on Decision and Control*, pages 4627–4632, San Diego, CA, USA, 2006.

J. Rudolph. Flachheit: Eine nützliche Eigenschaft auch für Systeme mit Totzeiten (Flatness: A useful property also for systems with delays). *at – Automatisierungstechnik*, 53(4–5): 178–188, 2005.

J. Rudolph and J. Winkler. A generalized flatness concept for nonlinear delay systems: motivation by chemical reactor models with constant or input dependent delays. *International Journal of Systems Science*, 34(8–9):529–541, 2003.

R. Scattolini. Architectures for distributed and hierarchical model predictive control – a review. *Journal of Process Control*, 19(5):723–731, 2009.

M. Schulze Darup and M. Mönnigmann. Explicit feasible initialization for nonlinear MPC with guaranteed stability. In *Proceedings of the 50th IEEE Conference on Decision and Control and European Control Conference*, pages 2674–2679, Orlando, FL, USA, 2011.

P. O. M. Scokaert, D. Q. Mayne, and J. B. Rawlings. Suboptimal model predictive control (feasibility implies stability). *IEEE Transactions on Automatic Control*, 44(3):648–654, 1999.

J. S. Shamma and D. Xiong. Linear nonquadratic optimal control. *IEEE Transactions on Automatic Control*, 42(6):875–879, 1997.

H. Shang, J. F. Forbes, and M. Guay. Computationally efficient model predictive control for convection dominated parabolic systems. *Journal of Process Control*, 17(4):379–386, 2007.

Y.-J. Shi, T.-Y. Chai, H. Wang, and C.-Y. Su. Delay-dependent robust model predictive control for time-delay systems with input constraints. In *Proceedings of the American Control Conference*, pages 4880–4885, St. Louis, MO, USA, 2009.

A. Sideris and J. E. Bobrow. An efficient sequential linear quadratic algorithm for solving nonlinear optimal control problems. *IEEE Transactions on Automatic Control*, 50(12): 2043–2047, 2005.

S. Summers, D. M. Raimondo, C. N. Jones, J. Lygeros, and M. Morari. Fast explicit nonlinear model predictive control via multiresolution function approximation with guaranteed stability. In *Proceedings of the 8th IFAC Symposium on Nonlinear Control Systems*, pages 533–538, Bologna, Italy, 2010.

P. L. Tang and C. W. De Silva. Stability validation of a constrained model predictive networked control system with future input buffering. *International Journal of Control*, 80(12):1954–1970, 2007.

A. R. Teel and L. Praly. Global stabilizability and observability imply semi-global stabilizability by output feedback. *Systems & Control Letters*, 22(5):313–325, 1994.

F. Tröltzsch and D. Wachsmuth. On convergence of a receding horizon method for parabolic boundary control. *Optimization Methods and Software*, 19(2):201–216, 2004.

S. E. Tuna, M. J. Messina, and A. R. Teel. Shorter horizons for model predictive control. In *Proceedings of the American Control Conference*, pages 863–868, Minneapolis, MN, USA, 2006.

T. Utz. *Control of Parabolic Partial Differential Equations Based on Semi-Discretizations*. PhD thesis, TU Wien, 2012.

T. Utz, K. Graichen, and A. Kugi. Trajectory planning and receding horizon tracking control of a quasilinear diffusion-convection-reaction system. In *Proceedings of the 8th IFAC Symposium on Nonlinear Control Systems*, pages 587–592, Bologna, Italy, 2010.

M. Vidyasagar. *Nonlinear Systems Analysis*. Prentice-Hall, Englewood Cliffs, NJ, USA, 2nd edition, 1993.

K. Worthmann. Estimates on the prediction horizon length in MPC. In *Proceedings of the 20th International Symposium on Mathematical Theory of Networks and Systems*, Melbourne, VIC, Australia, 2012a.

K. Worthmann. *Stability Analysis of Unconstrained Receding Horizon Control Schemes*. PhD thesis, University of Bayreuth, 2012b.

K. Worthmann, M. Reble, L. Grüne, and F. Allgöwer. The role of sampling for stability and performance in unconstrained nonlinear model predictive control. *SIAM Journal on Control and Optimization*, 2012. submitted.

L. Würth, J. B. Rawlings, and W. Marquardt. Economic dynamic real-time optimization and nonlinear model-predictive control on infinite horizons. In *International Symposium on Advanced Control of Chemical Processes*, Istanbul, Turkey, 2009.

T. H. Yang and E. Polak. Moving horizon control of nonlinear systems with input saturation, disturbances and plant uncertainty. *International Journal of Control*, 58(4):875–903, 1993.

S. Yu, M. Reble, H. Chen, and F. Allgöwer. Inherent robustness properties of quasi-infinite horizon MPC. In *Proceedings of the 18th IFAC World Congress*, pages 179–184, Milan, Italy, 2011.

S. Yu, M. Reble, H. Chen, and F. Allgöwer. Inherent robustness properties of quasi-infinite horizon nonlinear model predictive control. *Automatica*, 2012. submitted.

L. A. Zadeh and B. H. Whalen. On optimal control and linear programming. *IRE Transactions on Automatic Control*, 7(4):45–46, 1962.

M. N. Zeilinger. *Real-time Model Predictive Control*. PhD thesis, ETH Zürich, 2011.

M. N. Zeilinger, C. N. Jones, and M. Morari. Real-time suboptimal model predictive control using a combination of explicit MPC and online optimization. *IEEE Transactions on Automatic Control*, 56(7):1524–1534, 2011.

A. Zemouche, M. Boutayeb, and G. I. Bara. Observer design for a class of nonlinear time-delay systems. In *Proceedings of the American Control Conference*, pages 1676–1681, New York, NY, USA, 2007.

L. Zhilin, Z. Jun, and P. Run. Robust model predictive control of time-delay systems. In *Proceedings of the IEEE Conference Control Applications*, pages 470–473, Istanbul, Turkey, 2003.